Multi-Power
MICROWAVE MIRACLES

FROM

FROM THE LIBRARY OF
Sonia Ayotte

A Rutledge Book
The Benjamin Company, Inc.
New York, New York

by Hyla O'Connor

Library of Congress Catalog Card Number: 78-52221 ISBN: 0-87502-061-5
Prepared and produced by Rutledge Books.
Published by The Benjamin Company, Inc.
485 Madison Avenue
New York, New York 10022
Printed in U.S.A.
First printing
Second printing October 1978

Photographs by Walter Storck Studios, Inc.; photograph, page 25, by William Pell
Illustrations by Tom Huffman

Contents

MARVEL AT COOKING WITH MICROWAVES

Naturally enough, you're anxious to begin using your new microwave oven. But do take the time to review the material in this section. It's not only interesting and helpful; it will pay off for you in terms of getting the most out of your oven.

Although the way the microwave oven works may seem almost miraculous, the basic principles behind its cooking of food are not difficult to understand.

What Microwave Energy Is

The electrical energy for the microwave works on standard household current 110–120v. In the case of a microwave oven, the electrical energy is converted into electromagnetic energy by means of an electron tube located within the oven. This tube, called a magnetron, converts electrical energy into electromagnetic—or microwave—energy, then sends microwaves directly into the food to be cooked.

How Microwaves Cook Food

Microwaves, transmitted by the magnetron, bounce off the interior sides of the oven and pass through suitable cooking utensils to come in direct contact with the food. These microwaves—or short energy waves—are attracted to molecules of moisture, fat, and sugar in the food; they cause the molecules to rub against each other, and eventually to vibrate rapidly. This in turn sets up friction, and the resulting heat engendered cooks the food.

How the Oven Works

When you cook with a conventional range, food heats and is cooked by means of electricity, gas, or wood. On top of the range, the heat applied to the bottom of the pan cooks the food; when food is placed in an oven, the hot air surrounding the food cooks it.

Not so with the microwave oven. The microwaves travel directly to the food. They do not heat the surrounding air or the recommended dishes in which the food is being cooked. There is also no need to wait for the oven to heat, because the cooking action begins instantly.

Installation and Maintenance of the Oven

Both installation and maintenance are quite simple. To install your oven, *follow manufacturer's directions carefully*. A microwave oven operates on standard household current and does not require an expert to ready it for regular use.

Maintenance requires a few simple cleaning steps. Please *follow your manufacturer's directions*. Since there is little splattering with microwave cooking, you'll find that the buildup of grease or the like is minimal; an occasional wiping is all that is required to keep the oven clean. About the only other maintenance necessary is occasional replacement of the light bulb in the interior of the oven.

Keep the door and door gasket free of food buildup to maintain a tight seal.

Using the Oven

The recipes in this book refer to ten cooking levels: WARM, LOW, DEFROST, BRAISE, SIMMER, BAKE, ROAST, REHEAT, SAUTE, and HIGH. Each cook level corresponds to a certain microwave power level as indicated on the oven.

The multi-power settings available on your oven open up the possibility of almost infinite variations in cooking. The percentages of microwave power range from 10% to 100%. Each of the basic ten settings corresponds to a cooking process or technique. The chart below gives a good idea of what these processes are and what standard uses they can be put to.

COOKING GUIDE FOR MULTI-POWER SETTINGS

Setting	Power	Uses and Information
WARM	10%	Softening cream cheese; proofing bread dough; keeping dinners warm.
LOW	20%	Softening chocolate; heating breads, rolls, pancakes, and French toast.
DEFROST	30%	Defrosting frozen foods; cooking noodles and rice.
BRAISE	40%	Cooking less-tender cuts of meat in liquid (e. g., pot roast, Swiss steak).
SIMMER	50%	Cooking sauces, stews, and soups.
BAKE	60%	Starting cakes and quick breads (such as biscuits and corn breads).
ROAST	70%	Cooking rump roasts, ham, veal, and lamb; cooking cheese dishes; defrosting large cuts of meat and poultry.
REHEAT	80%	Reheating leftovers.
SAUTE	90%	Quickly frying onions, mushrooms, green peppers.
HIGH ("HI")	100%	Cooking poultry, fish, vegetables, and most casseroles; preheating the browning dish.

Because Multi-Power Cook Control allows you to choose any setting you want, you will be able to cook with great flexibility and precision. You can control results so that the end product suits your taste exactly.

The chart below is a guideline to cooking foods to doneness according to their internal temperature.

TEMPERATURE CONTROL

100°	Reheat Baby Bottle
110°	Baked Goods
120°	Fully Cooked Ham
130°	Rare Beef
140°	Medium Beef
150°	Well-Done Beef, Lamb, Reheat Casseroles
160°	Veal, Pork, Meat Loaf, Reheat Vegetables
170°	Reheat Beverages
180°	Simmer Soups
190°	Puddings

Utensils and Containers

Glass, china, and pottery utensils are ideal for use in a microwave oven because microwaves pass through them. If any of these have metallic trim or glaze, they should not be used.

If you are unsure about the suitability of a particular utensil, place it in the oven on the HIGH setting for 30 seconds. A container that remains cool is safe to use. If it is warm, don't use it.

Check manufacturer's labels, too. Often a utensil will be labeled "Good for Microwave."

Plastics are safe to use if they are dishwasher safe—but use them only for limited cooking periods or for heating. Do not use plastic for tomato-based foods or foods with a high fat or sugar content.

Plastic cooking pouches can be used provided that: 1) metal ties or twisters are removed first and 2) the pouch is slit so that steam can escape.

Paper is good for short-term cooking and for reheating on low heat. It cannot be foil-lined, and remember that prolonged use may cause it to burn. Waxed paper is suitable for preventing splattering.

Straw and wood can be used for quick warming—for instance, for heating rolls. Longer use could cause cracking.

A *browning dish* must be preheated; it's good for browning foods that have too short a cooking time to brown by themselves (pork chops, for instance).

A *microwave roasting rack* made of plastic is suitable;. it allows juices to drain from meat or poultry.

Most *metals* are not suitable for use in the microwave oven because they reflect microwaves and interfere with effective cooking. Exceptions to this rule are listed below. If any metal touches the sides of the oven it can cause arcing (sparks in the oven). Any metal used should be at least one inch from the oven wall; otherwise, arcing will occur and may damage the oven.

Exceptions to the No-Metal Rule:

On large pieces of meat or poultry, strips of aluminum foil can be used to cover areas that are becoming overcooked.

TV dinners can be heated in their shallow aluminum trays. (Metal containers used in the oven should not exceed three-fourths of an inch in depth.) However, TV dinners will heat much faster if you "pop" the blocks of food out and arrange them on ordinary dinner plates. (As a general rule of thumb, remember that food mass must always be greater than the amount of metal.)

Remove all metal twisters or foil strips from paper or plastic bags and substitute string; metal or foil may arc and ignite the paper or plastic.

It is very important that only microwave oven meat thermometers be used in the food during microwave cooking. Conventional meat thermometers may be inserted to test the internal temperature after the food has been removed from the oven. If the food has to be cooked further, remove the thermometer before you return the food to the microwave oven.

Do Not Use:

1) Metal twists or ties
2) Pots and pans
3) Baking sheets
4) Dishes with metallic trim or handles

All of these may cause a distorted cooking pattern, arcing, or both.

RECOMMENDED UTENSILS AND CONTAINERS

Item	Good Use	General Notes
China plates and cups (without metal trim)	Heating dinners and coffee	
Pottery plates, mugs, and bowls (unglazed)	Heating dinners, soups, and coffee	
Earthenware (ironstone) plates, bowls, and mugs	Heating dinners, soups, and coffee	If dish has been in refrigerator, may take longer to heat food.
Corelle® Livingware dinnerware	Heating dinners, soups, and coffee	Closed-handle cups should not be used.
Paper plates, cups, and napkins	Heating leftovers, coffee, frankfurters, doughnuts, and rolls	Absorb moisture from baked goods and freshen them.
Soft plastics, such as dessert topping cartons and Tupperware®	Reheating leftovers	Can be used for short reheating periods. Do not use to reheat acid-based foods or those with a high fat or sugar content.
Corningware® casseroles	Cooking main dishes, vegetables, and desserts	
Pyrex® casseroles	Cooking main dishes, vegetables, and desserts	Do not use dishes with metal trim, or arcing may occur.

Item	Good Use	General Notes
TV dinner trays (metal)	Frozen dinners or homemade dinners	Can be no deeper than ¾". However, microwaves will penetrate from the top and the food will receive heat from the top surface only.
Oven film and cooking bags	Cooking roasts or stews	Substitute string for metal twister (twister would cause arcing and bag would melt). Bag itself will not cause tenderizing. Do not use film with foil edges.
Cooking pouches	Cooking meats, vegetables, rice, and other frozen foods	Slit pouch so steam can escape.
Waxed paper	Wrapping corn on the cob; covering casseroles	Microwaves have no effect on wax. However, food temperature may cause some melting. (Wax will not adhere to hot food.)
Plastic wrap	Covering dishes	Puncture to allow steam to escape.
Metal spoons (not silver)	Stirring puddings and sauces	Will not cause arcing as long as there is a quantity of food. Handle may become warm. Do not leave in oven with small amounts of food.
Wooden spoons	Stirring puddings and sauces	Can withstand microwaves for short cooking periods.
Microwave roasting racks	Cooking roasts and chickens	
Browning dishes	Searing, grilling, and frying small meat items; grilling sandwiches	These utensils absorb microwaves and preheat to high temperatures. A special coating on the bottom makes them unique; they brown pieces of meat that otherwise would not brown in a microwave oven.

Important Terms in Microwave Cooking

There are a number of points that are important to remember for successful microwave cooking. Because the oven works so quickly, factors that would not be vital in conventional oven cooking become important. Following are some of the terms and cooking procedures that are integral to this new kind of cooking—along with an explanation of each.

Starting temperature of the food to be heated affects cooking time. Cooking time given in recipes is based on normal storage temperature of ingredients.

For example, milk is usually used right from the refrigerator. Therefore, any recipe that uses milk assumes that the milk will be cold, having been taken directly from the refrigerator.

If you use a particular ingredient that is colder than-normal, the cooking time will be a bit longer. Similarly, if an ingredient is warmer than would be usual, the cooking time will decrease to some extent.

In general, the warmer the food or ingredients to start with, the shorter the cooking or heating time. Remember this if you make substitutions in recipes. If you use a frozen ingredient instead of the canned one that the recipe specifies, you'll have to increase the cooking time given in the recipe.

Density refers to the composition of a particular food item. In other words, some foods have a basic structure that consists of molecules that are tightly packed together (a meat roast is one example). When this is the case, it takes microwaves longer to penetrate and cook the food than they would take for a less dense item. A more porous food (such as bread) absorbs microwaves faster and heats through or cooks more quickly.

Volume is a factor that affects cooking time in the microwave oven. Here the word volume refers to the amount of food to be cooked. If you are baking several potatoes, the cooking time will be longer than it would if you were baking just one potato. The same holds true with liquids to be heated: three cups of water will take longer than would just one cup.

If you halve a recipe, cook the food a little more than half the time the recipe calls for. In both cases, keep checking at short intervals until the food is cooked to your taste.

Arrangement of foods within the oven should be taken into consideration. If there are several pieces of similar food, arrange them so that each gets the maximum concentration of microwaves. Potatoes should be arranged in a ring; ears of corn are best placed like spokes of a wheel, from the center of the oven out to the sides. When you reheat a plate of leftovers, denser foods should be placed toward the outer edge of the container. The more porous foods (like bread or rolls) can be positioned in the center of the plate. This way, dense foods get the greatest concentration of microwave energy and there is a more even microwave distribution in all the foods.

Delicate ingredients require a lower setting for proper cooking. Many high-protein foods in the dairy group fall into this category: cheese, eggs, milk, cream, and sour cream. Cooking at a higher heat may cause these foods to toughen, separate, or curdle.

Mayonnaise, kidney beans, and mushrooms also require lower heat. On a higher setting, mayonnaise may separate, kidney beans and mushrooms may "pop."

All of the foods referred to above are penetrated quickly by microwaves and are easily overcooked. Using a lower setting guards against overcooking.

Container size and shape specified in recipes should be followed for best results.

If you vary the size or shape of the container, the cooking time may vary. A tall, narrow container will increase cooking time, just as a shallow, broad container will reduce it.

The containers specified in the recipes have been chosen for a reason. You will find that recipes for puddings and sauces call for containers that are larger than the quantity of liquid being cooked to prevent boilovers. Cake recipes call for round utensils for more even cooking. This is important to ensure good texture in a cake.

Coverings suitable for use in the microwave oven include glass covers, plastic wrap, waxed paper, and glass plates and saucers. Covers are useful because they trap steam and therefore speed cooking time. Furthermore, they seal in natural moisture, preventing foods from drying out as well as preserving nutrients.

Remove any covering away from hands and face to prevent steam burns.

Stirring is necessary for some foods. Because microwaves cook the outside edges of food first, the center portion sometimes needs redistribution for even cooking. This is true of puddings; the redistribution is accomplished simply by stirring. Always stir from the outside in, so that heat is equalized and uncooked portions flow toward the outside edges.

This technique is similar to the top-to-bottom stirring you're accustomed to doing on a range burner. In the microwave oven, however, you need stir only occasionally.

Browning usually takes place naturally. After ten minutes or so, meats and poultry brown. For individual pieces of meat that cook in a shorter period of time (such as hamburgers, pork chops, or steak), use a special microwave browning dish.

If you prefer, you can create a browned look by brushing on a gravy mix or bottled flavor enhancer after the meat is cooked.

Cookies, cakes, and breads do not brown well in the oven. When dark-colored ingredients (such as chocolate or spices) are part of the recipe, the lack of browning is not apparent. Other times, glazes or frostings can be used.

Turning foods over is sometimes necessary. In the case of large, dense foods (e.g., roasts), turning the food over will help to cook it evenly. Generally it is not necessary to rotate the food container. Chicken pieces, pork chops, and other meats with bones should be placed so that the bony part faces the center, the thick part faces the outside. This aids in even cooking.

Standing time is important to microwave cooking. The standing time specified in the recipes is really a part of the cooking time, in that food continues to cook after it is removed from the oven. The more dense the food, the longer the standing time. In addition to finishing the cooking process, standing time helps retain natural juices and makes carving easier.

The recipes in this book all take standing time into account and specify the correct amount of time to allow. With your own recipes, some experimentation may be necessary to gauge proper standing time. As a guideline, use the standing time specified in a similar recipe here.

Meal planning should not pose a problem. Keep in mind that foods that need to be cooked longest should be cooked first. The main dish (such as a roast or a casserole) should be cooked first, then the vegetables and bread cooked or heated. Desserts can be made in the morning—unless they are to be served hot.

Adapting Recipes

To get the most out of your microwave oven, your own favorite recipes can be adapted for microwave cooking.

First, find a recipe in this book that's similar to the recipe you'd like to try. Use the microwave recipe for guidelines on cooking time, container size, power setting, and ingredients. Or cook the dish for one-fourth of the conventional cooking time, then add time (in short segments) until the food is cooked to your liking.

This chart will give you a general idea of how to proceed.

GUIDELINES FOR ADAPTING YOUR OWN RECIPES

Foods	Suggested Setting(s)	General Information
Appetizers and Sandwiches	HIGH and ROAST	Use already-toasted bread. Do not assemble canapes until ready to microwave, or crackers and toast will become soggy. Appetizers with crust do not microwave well (pastry stays pale). Dips heat well, stay smooth, and do not scorch.
Rib, Leg of Lamb, Pork Loin Roasts	ROAST	Meat will brown somewhat but not as much as in a conventional oven. Standing time is especially important, as some cooking occurs after roast is removed from the oven.
Chuck and Round Roasts	BRAISE	Less browning than in a conventional oven. The LOW setting allows tough meat fibers to become tender, the longer the cooking time, the more tender the meat becomes. If you wish, you can pre-brown the roast on top of a conventional range. Cover, add liquid.

Foods	Suggested Setting(s)	General Information
Stewing Beef	BRAISE	Depending on cut and size of pieces, will cook at different rates. Acceptable results in minimum time, but the longer the cooking time, the more tender the meat will be. Use less liquid than you would in a conventional recipe.
Bacon, Steak, Chops	HIGH	Bacon will brown because of its high fat content. Use browning dish for steaks and chops.
Meat Loaf and Ham	ROAST	Browning approximately the same as with conventional cooking. Cured meats contain sugar, concentrated in spots, and may overcook if not watched carefully.
Poultry and Game Birds	HIGH and ROAST	Poultry becomes very tender. Skin will be soft except for more fatty birds, such as ducks. Final color is golden brown rather than crispy brown.
Fish and Seafood	HIGH	Fish will retain more moisture than in conventional cooking. Remains tender; cooking in a sauce ensures excellent results.

Foods	Suggested Setting(s)	General Information
Eggs and Custard	ROAST	Do not cook eggs in the shell (they will explode). Scrambled eggs are light and tender. Fried eggs can be cooked in browning dish. Do not cook puffy omelets. Custard requires ROAST setting to avoid curdling.
Cheese	ROAST	Cheese should be melted or cooked at a ROAST setting. Cheese sauce and fondue should be stirred occasionally.
Rice and Pastas	DEFROST	Microwave saves some time, but not a great deal. Add 1 tablespoon of salad oil to boiling water to prevent boilovers. Use large dish.
Fruits and Vegetables	HIGH	Tender-crisp results. Very little additional water needed, with maximum amount of natural moisture retained. No scorching.

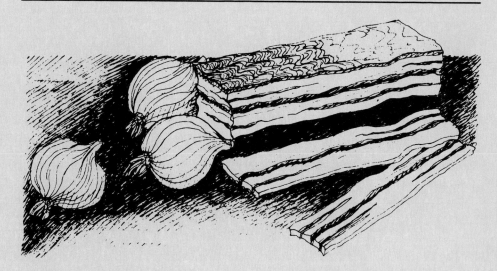

Foods	Suggested Setting(s)	General Information
Cakes, Quick Breads, Yeast Breads, Cookies	BAKE	These do not form a crust, are lighter than conventionally baked goods. Top of cakes may be wet-looking after cooking. Do not overcook to remove the moist appearance, or product will toughen. When you adapt a conventional recipe, reduce baking powder and soda by a fourth or a third. Baked products will be pale (except chocolate or spice mixtures). Angel food and chiffon cakes do not bake well in the microwave. Microwave works best for bar cookies that need no browning and should remain soft.
Frostings and Candies	ROAST	These sugar mixtures cook very quickly, with excellent results. Use buttered large heat-proof bowl. Very little stirring required. Check temperature with candy thermometer after cooking periods (and after removal from oven).
Sauces and Fillings (thickened by flour or cornstarch)	HIGH	Microwave gives excellent results. Blend flour or cornstarch well before cooking. Stir halfway through cooking period in order to prevent lumping. Use slightly less liquid than in conventional recipes. Will not scorch.

Foods	Suggested Setting(s)	General Information
Pies	HIGH and ROAST	Crust becomes flaky but does not brown. Cook crust first before adding filling. Excellent results are achieved by starting in the microwave and finishing in a conventional oven.
Frozen Foods	REHEAT, ROAST, and SIMMER	Reheat in microwave. Cover foods to retain moisture. Use a container that conforms to the shape of the frozen food.

Many kinds of dishes—but nothing metal—can be used in the microwave oven.

USER INSTRUCTIONS

PRECAUTIONS TO AVOID POSSIBLE EXPOSURE TO EXCESSIVE MICROWAVE ENERGY

(a) DO NOT ATTEMPT to operate this oven with the door open since open-door operation can result in harmful exposure to microwave energy. It is important not to defeat or tamper with the safety interlocks.

(b) DO NOT PLACE any object between the oven front face and the door or allow soil or cleaner residue to accumulate on sealing surfaces.

(c) DO NOT OPERATE the oven if it is damaged. It is particularly important that the oven door closes properly and that there is no damage to the:
   (1) DOOR (bent)
   (2) HINGES AND LATCHES (broken or loosened)
   (3) DOOR SEALS AND SEALING SURFACES.

(d) THE OVEN SHOULD NOT BE ADJUSTED OR REPAIRED BY ANYONE EXCEPT PROPERLY QUALIFIED SERVICE PERSONNEL.

Appetizers

Piping hot appetizers, prepared days or weeks in advance and frozen, can be served in minutes with the aid of the microwave oven. Canapés with a bread or cracker base should be assembled just before warming to prevent sogginess. Canapés should be placed on a tray or platter with an underliner of paper napkins or towels to absorb any moisture. When serving the heated canapés, discard the paper underliner and place the canapés directly on the serving platter—not merely for the sake of attractive service, but also to keep the canapés from resting on a damp surface.

The microwave oven is ideal for dips and dunks of all kinds. They can be prepared long in advance, placed in serving bowls, and heated in the serving dish with a minimum of time and bother.

Cocktail Wieners

6 to 8 servings

¼ cup minced onion

2 teaspoons butter

½ cup catsup

1 tablespoon vinegar

½ teaspoon Worcestershire sauce

2 tablespoons brown sugar

½ teaspoon salt

½ teaspoon dry mustard

½ teaspoon paprika

2 to 3 dozen cocktail wieners

1. Combine onion and butter in a 1- to 1½-quart casserole.
2. Cook, covered, on SAUTE for 3 minutes, or until onion is soft.
3. Add ¼ cup water and stir in remaining ingredients except cocktail wieners.
4. Cook, covered, on HIGH for 2½ minutes, or until sauce is bubbly.
5. Stir in cocktail wieners.
6. Cook, covered, on HIGH for 4 minutes, or until wieners are hot.
7. Let stand, covered, for 2 minutes.
8. Serve warm, use toothpicks to spear individual wieners.

Bean Dip

3 cups

1 can (1 pound) baked or kidney
  beans

1 jar (8 ounces) pasteurized process
  cheese spread

¼ cup chili sauce

1 teaspoon chili powder

Dash of hot-pepper sauce

1. Pour beans into a 1½-quart casserole. Mash beans with a fork. Add remaining ingredients and blend well.
2. Cook, covered, on BAKE for 2 to 3 minutes. Stir well.
3. Cook, covered, on SIMMER for 2 to 3 minutes, or until piping hot.
4. Keep hot on a heated tray or a candle warmer.

Hot Mexican Cheese Dip

2½ cups

1 pound process American cheese,
  grated

1 can (10 ounces) green chilies and
  tomatoes

1. Combine cheese and chilies in a 1½-quart casserole.
2. Cook, covered, on REHEAT for 5 minutes. Stir thoroughly.
3. Keep hot on a heated tray or candle warmer, and serve with corn chips or potato chips or with crisp raw vegetables.

Sweet and Sour Hot Dogs

6 to 8 servings

2 tablespoons prepared mustard

¼ cup grape jelly

½ pound frankfurters

1 teaspoon butter or margarine, melted

1. Combine mustard and jelly in a 1-cup measuring cup.
2. Cook, covered with plastic wrap, on BAKE for 3 minutes.
3. Cut each frankfurter diagonally in 9 to 10 slices. Put in a 1-quart baking dish with butter.
4. Cook, covered, on HIGH for 2 minutes.
5. Pour grape jelly sauce over frankfurters. Cook, covered, on HIGH for 4 minutes.
6. Let stand 2 minutes.
7. Serve hot; use toothpicks to pick up individual slices.

Tantalizer Spread

¾ cup

3 strips bacon

1 small tomato, peeled and quartered

1 teaspoon prepared mustard

1 package (3 ounces) cream cheese, cut into cubes

¼ teaspoon celery salt

½ cup blanched almonds

1. Place paper towels in bottom of an 8-inch square dish. Place bacon strips on towels. Cover loosely with another paper towel.
2. Cook on HIGH for 3½ minutes, or until bacon is very crisp.
3. Put tomato, mustard, cream cheese, and celery salt in blender container. Cover and process until mixture is smooth. Add almonds and bacon and process only until almonds are chopped.
4. Serve with an assortment of crackers.

Cheddar Cheese Canapés

24 canapés

¼ cup grated Cheddar cheese

2 tablespoons cream

1 tablespoon grated Parmesan cheese

⅛ teaspoon Worcestershire sauce

⅛ teaspoon hot-pepper sauce

1 tablespoon sesame seeds

24 rounds of toast or crisp crackers

Chopped parsley

1. Combine Cheddar cheese, cream, Parmesan cheese, Worcestershire, hot pepper sauce, and sesame seeds. Blend with an electric mixer until smooth.
2. Spread 1 teaspoon of the mixture on each of the toast rounds or crackers.
3. Arrange 12 crackers on a platter lined with paper towels.
4. Cook on ROAST for 20 to 30 seconds, or until mixture is warm and cheese is melted. Repeat with remaining crackers.
5. Garnish with parsley; serve warm.

Crab Supremes

16 to 18 canapés

1 can (6½ to 7 ounces) crab meat
½ cup finely minced celery
2 teaspoons prepared mustard
4 teaspoons chopped sweet
  pickle relish

½ cup mayonnaise
Crisp crackers or toast rounds

1. Drain crab meat. Place in a 1-quart bowl and flake with a fork. Add celery, mustard, pickle relish, and mayonnaise. Mix well.
2. Spread mixture on crackers or toast rounds. Place 8 at a time on a plate lined with a paper towel. Cover with waxed paper.
3. Cook on ROAST for 30 to 45 seconds, or until piping hot.

Shrimp Olive Dip

about 2½ cups

1 can (10½ ounces) cream of shrimp
  soup, undiluted
1 package (8 ounces) cream cheese,
  cut into chunks
1 can (8 ounces) chopped ripe olives,
  drained

2 tablespoons lemon juice
1 teaspoon Worcestershire sauce
¾ teaspoon curry powder
  (optional)

1. Combine soup and cream cheese in a 1-quart casserole.
2. Cook, covered, on BAKE for 3 minutes.
3. Remove and stir until cheese is well blended. Stir in remaining ingredients.
4. Cook, covered, on BAKE for 2 to 3 minutes, or until piping hot.
5. Place over a candle warmer to keep hot and serve as a dip with corn chips or potato chips.

Stuffed Mushrooms

makes 8 to 10

8 to 10 medium mushrooms
2 tablespoons minced onion
3 tablespoons butter or margarine

¼ cup dry bread crumbs
¼ teaspoon hot-pepper sauce
2 tablespoons sherry

1. Wipe mushrooms with a damp paper towel. Carefully twist off stems from mushroom caps, leaving caps intact. Chop stems very fine. Place in a small mixing bowl with onion and butter. Cook, uncovered, at HIGH for 3 to 4 minutes, or until onion is tender.
2. Combine mushroom mixture with bread crumbs, hot-pepper sauce, and sherry and mix well. Fill mushroom caps with this stuffing. Place filled mushrooms in an 8-inch baking dish.
3. Cook, covered, on BAKE for 4 minutes, or until mushrooms are piping hot.

11

Meatball Appetizers

about 60

1 pound ground beef
½ pound ground pork
1 small onion, finely minced
1 cup milk
1 egg, lightly beaten

1 cup dry bread crumbs
1 teaspoon salt
¼ teaspoon pepper
¼ teaspoon ground allspice

1. Combine ingredients in a large mixing bowl and blend well. Form into small balls, about 1 inch in diameter.
2. Arrange half of the meatballs in a single layer in an oblong baking dish.
3. Cook, uncovered, on BAKE for 4 minutes.
4. Place in a chafing dish to keep hot.
5. Cook remaining meatballs.
6. Serve hot with toothpicks and favorite dunking sauce, if desired.

Shrimp Dip

2 cups

1 can (10½ ounces) cream of shrimp soup, undiluted
1 package (8 ounces) cream cheese, softened

1 teaspoon lemon juice
Dash of paprika
Dash of garlic powder

1. Pour soup into a small mixing bowl.
2. Cook, covered, on BAKE for 3 minutes, or until hot. Stir well.
3. Beat in cream cheese, lemon juice, paprika, and garlic powder. Cook on BAKE for 1½ to 2 minutes, or until piping hot.
4. Keep warm over a candle warmer. Serve as a dip with crisp vegetables.

Cocktail Shrimp

2 to 3 servings

½ pound medium raw shrimp
¼ cup butter or margarine

1 clove garlic, minced
3 tablespoons dry white wine

1. Remove shells and veins from shrimp. Rinse in cold water.
2. Place butter, garlic, and white wine in an 8-inch baking dish or shallow casserole.
3. Cook on HIGH for 2 minutes.
4. Stir mixture and place shrimp in dish. Cover loosely with a piece of waxed paper.
5. Cook on HIGH for 4 minutes, or until shrimp are pink in color and just tender. Do not overcook or shrimp will be tough.
6. Pour shrimp and sauce into a small bowl. Season with salt.
7. Chill and serve as an appetizer with desired dunking sauce.

Rumaki

about 36 appetizers

1 can (8 ounces) water chestnuts   **12 slices bacon**

1. Drain water chestnuts and cut into thirds.
2. Cut bacon slices into thirds. Place a piece of water chestnut on a piece of bacon and roll bacon around it. Secure with a toothpick. Roll remaining water chestnuts and bacon.
3. Place 10 at one time on a plate covered with paper towels. Cover top with a paper towel. Cook on HIGH for 2 minutes. Turn over. Cook on HIGH for 1 to 2 more minutes, or until bacon is cooked.
4. Let stand 1 minute before serving.

Mushroom-Egg Canapés

18 canapés

2 tablespoons butter or margarine	1 tablespoon chopped parsley
1 tablespoon finely minced onion	½ teaspoon salt
½ cup finely chopped mushrooms	18 toast rounds or crackers
2 hard-cooked eggs, finely chopped	¼ cup grated Cheddar cheese

1. Put butter, onion, and mushrooms in a small mixing bowl. Cook, uncovered on SAUTE for 3 minutes, or just until mushrooms are soft.
2. Stir in eggs, parsley, and salt.
3. Spread 1 rounded teaspoon of mixture on each toast round. Sprinkle with grated Cheddar cheese.
4. Arrange on a tray lined with paper towels.
5. Cook on REHEAT for 30 seconds, or just long enough to melt the cheese.
6. Serve warm.

Deviled Toasties

30 to 35 canapés

½ pound lean ground round steak	½ teaspoon prepared mustard
2 tablespoons minced onion	⅛ teaspoon prepared horseradish
1 teaspoon catsup	1 loaf party rye bread

1. Combine meat, onion, catsup, mustard, and horseradish. Blend well. Put about 1 teaspoon of the mixture on the top of each little slice of bread.
2. Cook, uncovered, 9 at a time on a paper towel on HIGH for 50 seconds, or to the desired degree of doneness. Remove to serving plate at once.

Soups

The microwave oven makes the preparation of soups, stews, and chowders as easy as one, two, three, and as individual as each person at the table. Prepared soups can be heated in glass or pottery serving bowls—perfect for quick service, with never a pot to wash. Containers should hold at least double the volume of soup they contain. Most soups are cooked covered with glass lids or plastic wrap. They should be stirred as specified for even distribution of ingredients.

The settings recommended are: for soups made with raw vegetables, fish, or seafood, use HIGH; uncooked meat or chicken soups, start on HIGH, finish on SIMMER; cooked meats and/or vegetables, use REHEAT; soups with milk, cream, or clams, use ROAST.

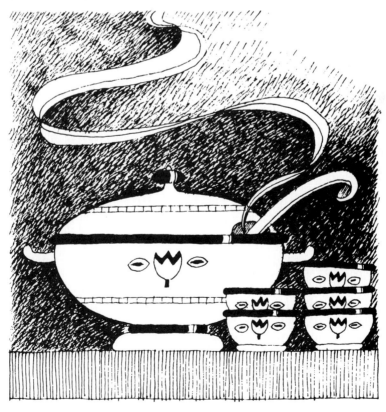

Quebec Green Pea Soup

4 to 6 servings

1 can (2 ounces) mushroom stems and pieces

1 tablespoon butter or margarine

2 cans (11½ ounces) condensed green pea soup, undiluted

1 cup grated raw carrots

1. Drain mushrooms, pouring liquid into a measuring cup. Add enough water to make 2 cups of liquid.
2. Melt butter in a 1½- to 2-quart casserole on BAKE for 30 seconds. Add drained mushrooms.
3. Cook on HIGH for 3 minutes, or just until heated.
4. Add soup and mushroom-water mixture. Stir until well blended. Stir in grated carrots.
5. Cook, covered, on ROAST for 7 to 9 minutes, or just until carrots are crisply tender.
6. Taste and season if necessary. Serve with croutons or crackers.

Spring Pea Soup

4 servings

1 package (10 ounces) frozen peas

3 cups chicken broth

6 scallions, sliced

¼ teaspoon white pepper

½ cup heavy cream

1. Place peas in a 1½-quart ovenproof casserole. Add chicken broth, scallions, and pepper.
2. Cook, covered, on HIGH for 7 minutes, stirring once.
3. Stir in cream, taste, and add salt if necessary.
4. Cook on BAKE for 1 minute.

Note: If desired, after Step 2, puree soup in a blender or force through a sieve. Proceed with Step 3.

Chili Chowder

6 servings

¾ pound ground beef

1 medium onion, chopped

1 clove garlic, chopped

2 tablespoons chopped green pepper

1 can (1 pound) peeled plum tomatoes

2 cups tomato juice

1 teaspoon salt

⅛ teaspoon sugar

2 teaspoons chili powder, or to taste

1. Place beef, onion, garlic, and green pepper in a 2-quart ovenproof casserole.
2. Cook, covered, on HIGH for 4 minutes.
3. Remove casserole and stir contents to break up beef.
4. Add tomatoes, including liquid from can, stirring to break up tomatoes.
5. Add remaining ingredients and mix well.
6. Cook, covered, on HIGH for 7 minutes, or until piping hot.

Cream of Mushroom Soup

6 servings

2 cups chopped fresh mushrooms
½ teaspoon onion powder
⅛ teaspoon garlic powder
⅛ teaspoon white pepper

¼ teaspoon salt
2½ cups chicken broth
1 cup heavy cream

1. Combine mushrooms, seasonings, and broth in a 2-quart casserole.
2. Cook on HIGH for 4 minutes, stirring once.
3. Stir in cream.
4. Cook on BAKE for 2 minutes, or until piping hot.

Superb Cream of Chicken

3 servings

1 can (10½ ounces) condensed cream
   of chicken soup, undiluted
1 soup can milk

1 pimiento, diced
¼ cup chopped ripe olives
½ teaspoon turmeric

1. Combine all ingredients in a 1-quart ovenproof bowl or measuring cup.
2. Cook on HIGH for 3 to 4 minutes, or until piping hot.

Cream Sudanese

4 servings

1 can (10½ ounces) condensed cream
   of tomato soup, undiluted
1 can (10½ ounces) condensed pea
   soup, undiluted

½ cup heavy cream
3 tablespoons sherry

1. Place both soups and 1½ soup cans of water in a 1½-quart ovenproof bowl. Stir to blend well.
2. Cook on HIGH for 2 minutes.
3. Stir in cream and sherry.
4. Cook on BAKE for 2 minutes, or until piping hot.

Egg Drop Soup

4 servings

2 cans (13¾ ounces) chicken broth
1 tablespoon cornstarch
1 can (4 ounces) water chestnuts, diced
2 scallions, chopped, including green
   tops

2 eggs, slightly beaten
Salt to taste

1. Put chicken broth in a 1½-quart casserole or mixing bowl.
2. Cook, covered, on HIGH for 4 minutes.
3. Combine cornstarch with 2 tablespoons water. Stir in hot broth. Stir in water chestnuts and scallions.
4. Cook, covered, on HIGH for 2 minutes, or until mixture is clear and piping hot.
5. Remove and quickly stir in beaten eggs. Taste, and season with salt, if necessary.

Indienne Cream

6 servings

2 cans (10½ ounces each) condensed
  cream of celery soup, undiluted
2 soup cans milk

1 teaspoon curry powder
2 medium apples, peeled, cored, and
  diced

1. Place soup in a 2-quart ovenproof bowl. Stir in milk and curry powder, mixing well.
2. Cook on HIGH for 3 minutes, or until piping hot.
3. Stir in apples. Cook on HIGH for 1 minute.

Potato Parsley Soup

4 servings

3 cups peeled, diced potatoes
¼ cup chopped onion
¼ teaspoon salt
1 can (13¾ ounces) chicken broth

1 small bunch parsley, chopped
2 tablespoons cornstarch
1½ to 2 cups milk

1. Combine potatoes, onion, salt, and broth in a 2-quart casserole. Add parsley.
2. Cook, covered, on HIGH for 14 minutes, or until potatoes are tender.
3. Combine cornstarch with a small amount of cold milk. Stir into potato mixture. Add remaining milk.
4. Cook, uncovered, on HIGH for 3 to 4 minutes, or until mixture comes to a boil and is piping hot. Stir once during cooking time.

Cheese Soup

4 servings

2½ cups beef broth
½ cup chopped onions

½ teaspoon celery powder
1½ cups grated sharp Cheddar cheese

1. In a 2-quart bowl combine beef broth, onion, and celery powder.
2. Cook on HIGH for 3 minutes.
3. Stir in cheese, blending well.
4. Cook on BAKE for 1 minute, or until cheese is melted.
5. Stir well before serving.

Cold Cucumber Soup

4 servings

2 cups chicken broth
3 large cucumbers
½ teaspoon salt

¼ teaspoon white pepper
1 tablespoon grated onion
1 cup light cream

1. Place chicken broth in a 2-quart ovenproof bowl.
2. Peel and dice 2 of the cucumbers and half of the third. Add to the chicken broth along with salt, pepper, and onion.
3. Cook on HIGH for 14 minutes, or until cucumbers soften.
4. Remove from oven and puree in a blender or force through a sieve. Cool.
5. Stir in cream and refrigerate.
6. Just before serving, float several slices of cucumber on top of each serving.

Cioppino   8 servings

1 large onion, chopped
1 medium green pepper, seeded and
  chopped
½ cup thinly sliced celery
3 cloves garlic, minced
3 tablespoons olive oil
1 can (3 pounds 3 ounces) peeled
  Italian tomatoes with puree
1 can (8 ounces) tomato sauce
1 teaspoon basil
1 bay leaf

1 teaspoon salt
¼ teaspoon pepper
1 pound firm white fish
1 dozen mussels or littleneck clams
  in the shell
1½ cups dry white wine
½ pound whole shrimp, cleaned and
  deveined
½ pound scallops
Chopped parsley

1. Combine onion, pepper, celery, garlic, and olive oil in a 4-quart casserole.
2. Cook on SAUTE for about 5 minutes, or until onion is soft.
3. Mash tomatoes with a fork or potato masher so that whole tomatoes are broken up in small pieces. Add tomatoes to casserole. Add tomato sauce, basil, bay leaf, salt, and pepper.
4. Cook, covered, on HIGH for 15 minutes to blend flavors.
5. While sauce is cooking, cut white fish into serving pieces. Using a stiff brush, thoroughly scrub the mussels, cutting off their beards, or soak clams in cold water to which a little cornmeal has been added and then scrub under cold running water to remove any residue of mud and sand. Stir wine into tomato mixture. Add white fish, shrimp, and scallops.
6. Cook, covered, on HIGH for 10 minutes.
7. Place mussels or clams in a layer on top of fish in casserole.
8. Cook, covered, on HIGH for 10 minutes, or until shells are fully opened.
9. Discard any mussels or clams that are unopened.
10. Ladle soup into soup plates. Sprinkle with parsley and serve piping hot with French bread.

Turkey Stock   2 quarts

1 turkey carcass from a 10-pound turkey
  (approximately)
2 stalks celery with leaves
1 small onion

1 teaspoon salt
½ teaspoon peppercorns
Pinch of mixed herbs

1. Strip all the meat from the turkey carcass. Break up body bones and place with leg and wing bones in a 3- or 4-quart casserole. Add remaining ingredients. Fill casserole a little over half full with water.
2. Cook, uncovered, on HIGH for 45 minutes.
3. Strain and use stock as desired.

Carrot Chowder

5 to 6 servings

4 slices bacon
1 can (1 pound) diced carrots
1 tablespoon grated onion

¼ cup finely diced celery
2 cups chicken broth

1. Cook bacon according to instructions on page 60. Crumble and reserve.
2. Place carrots, including liquid from can, in a 2-quart ovenproof bowl. Add remaining ingredients and stir well.
3. Cook on HIGH for 3 minutes, or until piping hot.
4. Stir in crumbled bacon before serving.

Note: Three hard-cooked eggs peeled and chopped may be added to the chowder along with the bacon if desired.

Clam Chowder

3 to 4 servings

2 slices bacon
1 can (7 ounces) minced clams, with liquid
1 large potato, peeled and cubed

¼ cup minced onion
1 can (13 ounces) evaporated milk
Salt and pepper to taste
1 tablespoon butter

1. Put bacon slices in a 2-quart casserole. Cover with a piece of paper towel.
2. Cook on HIGH for 3 minutes, or until bacon is crisp.
3. Remove paper towel and bacon, leaving drippings in casserole. Crumble bacon into bits and reserve. Add clams, clam liquid, potato, onion, and ½ cup water to casserole.
4. Cook, covered, on HIGH for 9 minutes, or until potatoes are tender.
5. Add milk, crumbled bacon, salt and pepper to taste, and butter.
6. Cook, covered, on HIGH for 3 minutes, or just until mixture comes to a boil.
7. Let stand 2 minutes. Serve with crumbled common crackers if desired.

Red Bean Soup

6 servings

8 slices bacon
2 tablespoons bacon drippings
1 large onion, diced
1 clove garlic, crushed

2 cans (27 ounces each) red kidney beans
½ teaspoon salt
1 can (8 ounces) tomato sauce

1. Cook bacon according to directions on page 60. Crumble and reserve.
2. Place bacon drippings in a 2-quart casserole. Add onion and garlic and cook, covered, on HIGH for 2 minutes.
3. Add ½ cup water and remaining ingredients, including liquid from the canned beans. Mix well.
4. Cook, covered, on HIGH for 5 to 6 minutes to blend flavors.
5. Puree soup in a blender, half at a time, and return to casserole. If soup is too thick, stir in enough water to make it the desired consistency.
6. Cook on BAKE for 3 minutes, or until piping hot.
7. Stir in reserved bacon before serving.

Meats

Roasts, chops, hamburgers, and small cuts of tender meat cook beautifully in the microwave oven. Meats can be thawed in the original wrapping in the oven. You must, however, remove the bag twisters or staples before thawing. Place the food to be thawed in a flat glass dish to catch the drippings.

Cook tender beef roasts such as sirloin tip and rib on HIGH for first half of cooking time, and finish cooking on ROAST. Less tender cuts of meat are started on ROAST, then cooked on simmer. The meat should be removed at the recommended "lower-than-done" temperature, as it continues to cook after being removed from the oven. Let it stand until the desired degree of doneness is reached. If the meat does not reach the temperature desired, it is a simple matter to return it to the microwave oven for a few minutes longer.

Meat—How to Defrost

1. Defrost in original wrapping. Remove any metal twist-ties or clamps.
2. Place meat in glass baking dish.
3. Meat is thawed on the DEFROST setting, unless otherwise stated.
4. Meat will be cool in the center when removed from the oven. Meat will continue to defrost as it is prepared for cooking. Start with minimum time to prevent meat from cooking on outer edges. Adjusting meat will speed time.

Meat	Minutes per Pound	Setting	Standing Time	Special Notes
Beef				
Ground beef	5 to 6	DEFROST	5 minutes	Freeze in flat squares. Turn over once.
Ground beef patty	1 per patty	DEFROST	2 minutes	Defrost on plate.
Pot roast, chuck:				
4 pounds and under	3 to 5	DEFROST	10 minutes	Turn over once.
over 4 pounds	3 to 5	ROAST	10 minutes	Turn over twice.
Rib roast, rolled:				
3 to 4 pounds	6 to 8	DEFROST	30 to 45 minutes	Turn over once.
6 to 8 pounds	6 to 8	ROAST	1½ hours	Turn over twice.
Rib roast, bone in	5 to 6	ROAST	45 minutes to 1 hour	Turn over twice.
Rump roast:				
3 to 4 pounds	3 to 5	DEFROST	30 minutes	Turn over once.
6 to 7 pounds	3 to 5	ROAST	45 minutes	Turn over twice.
Round steak	4 to 5	DEFROST	5 to 10 minutes	Turn over once.
Flank steak	4 to 5	DEFROST	5 to 10 minutes	Turn over once.
Sirloin steak, ½ inch thick	4 to 5	DEFROST	5 to 10 minutes	Turn over once.
Lamb				
Cubed	7 to 8	DEFROST	5 minutes	Break up cubes halfway through defrosting time.
Roast:				
4 pounds and under	3 to 5	DEFROST	30 to 45 minutes	Turn over once.
over 4 pounds	3 to 5	ROAST	30 to 45 minutes	Turn over twice.
Pork				
Chops	4 to 6	DEFROST	5 to 10 minutes	Separate chops halfway through defrosting time. Add 1 more minute if needed.
Spareribs	5 to 7	DEFROST	10 minutes	
Roast:				
4 pounds and under	4 to 5	DEFROST	30 to 45 minutes	Turn over once.
over 4 pounds	4 to 5	ROAST	30 to 45 minutes	Turn over twice.

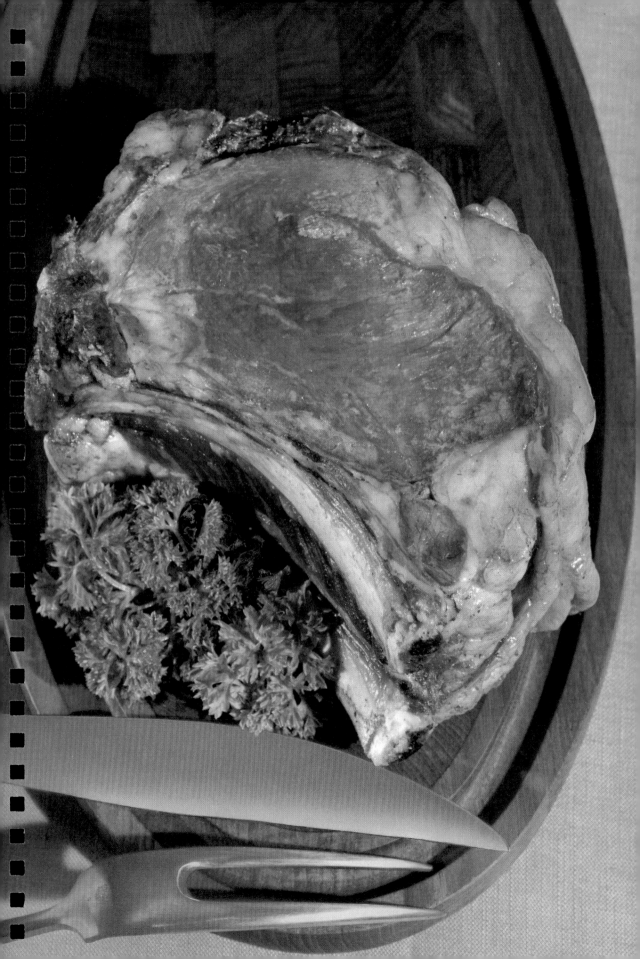

Meat	Minutes per Pound	Setting	Standing Time	Special Notes
Veal				
Roast:				
3 to 4 pounds	5 to 7	DEFROST	30 minutes	Turn over once.
6 to 7 pounds	5 to 7	ROAST	1 hour	Turn over twice.
Chops	4 to 6	DEFROST	20 minutes	Turn over once. Separate chops and continue defrosting.

Meat—How to Cook

1. Cook fresh or completely thawed meat.
2. Place meat, fat side down, on a microwave roasting rack in a glass baking dish. Use an inverted saucer if you do not have a rack, to keep the meat from resting in its juices. Turn meat over halfway through cooking time.
3. Meat that is to be roasted should be cooked uncovered; however, if the cut is fatty, a piece of waxed paper placed lightly over the top will keep the fat from splattering.
4. Large cuts of meat may have areas that cook faster than others. These can be covered with small pieces of aluminum foil to slow down cooking.
5. Use a temperature probe for accurate roasting of all cuts of meat. After turning meat over, select second setting and desired temperature. Cover tightly with aluminum foil when meat is removed from oven. During this standing time the internal temperature will rise approximately 15°F. A microwave meat thermometer may be used in the oven during cooking time, if desired, but *never* use a conventional meat thermometer. By inserting a conventional thermometer when meat comes out of the oven, the internal temperature of the meat can be checked.
6. Ground meats for casseroles or main dishes can be crumbled into a flat dish and cooked, covered with waxed paper or a glass lid.
7. Less tender cuts of meat require moist, slow cooking and should be tightly covered with a lid or plastic wrap. Close attention should be paid to slow-cooked dishes such as pot roast, stews, and swiss steak to be sure that the liquid does not evaporate.

Meat	First Setting	Second Setting	Minutes Per Pound	Meat Probe Temperature
Beef				
Ground Beef	HIGH		5	
Ground beef patty*	HIGH 2½ to 3 minutes Turn once 5 to 5½ minutes			

*Use of browning dish may be desired.

26

Meat	First Setting	Second Setting	Minutes per Pound	Meat Probe Temperature
Beef				
Meat loaf, 2 pounds	ROAST 25 to 30 minutes			160° F.
Rib roast, rolled: 3 to 4 pounds	HIGH	ROAST	Rare: 8 to 9 Medium: 10 to 11 Well-done: 12 to 13	130° F. 140° F. 150° F.
6 to 8 pounds	HIGH	ROAST	Same as above	
Rib roast, bone in	HIGH	ROAST	Rare: 7 to 8 Medium: 9 to 10 Well-done: 11 to 12	130° F. 140° F. 150° F.
Rump roast or chuck pot roast (cook covered and in a liquid)	HIGH 5 minutes per pound	SIMMER 10 minutes per pound	15 to 17	
Round steak (in liquid)	HIGH 5 minutes per pound	SIMMER 15 minutes per pound	20	
Sirloin steak*	HIGH		4½	
Lamb				
Cubed (in liquid)	Follow lamb recipes			
Roast	ROAST		10 to 12	150° F.
Pork				
Chops*	Check recipe section			
Spareribs	ROAST		12 to 15	
Roast	HIGH	ROAST	10 to 12	160°F.
Ham: Boneless: ready-to-eat	ROAST		11 to 13	120° F.
Shank of Leg	ROAST		8 to 9	120° F.
Canned: 3-pound	ROAST		10 to 11	120° F.
5 pound	ROAST		8 to 9	120° F.
Veal				
Roast	ROAST		18 to 20	160°F.
Chops	Check recipe section			

*Use of browning dish may be desired.

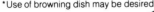

Pot Roast in Wine

about 6 servings

1 3-pound beef chuck pot roast
½ cup dry red wine
2 large onions, thinly sliced

3 medium potatoes, cut in quarters
4 carrots, peeled and sliced
Salt and pepper

1. Place pot roast in a 3-quart casserole or baking dish. Add wine and cover with sliced onions. Cook, covered, on HIGH for 5 minutes.
2. Cook on SIMMER for 1½ hours. If liquid is low, add water.
3. Add potatoes, carrots, and salt and pepper. Cook, covered, on SIMMER for 45 minutes, or until vegetables and meat are tender.
4. Let stand, covered, 5 to 10 minutes before serving.

Onion Steak

4 servings

¼ cup all-purpose flour
1 teaspoon salt
¼ teaspoon pepper

1½ to 2 pounds round steak
½ package (1½ ounces) onion soup mix

1. Combine flour, salt, and pepper. Place steak on a board and pound half of the flour mixture into each side of the steak with the back of a heavy knife. Cut meat in 4 pieces and place in an 8-inch square glass baking dish. Sprinkle any remaining flour over meat. Combine onion soup mix with 1 cup water. Pour over meat.
2. Cook, covered, on HIGH for 5 minutes. Cook on BAKE for 20 minutes. Turn meat over. Cook, covered, on BAKE for 20 minutes, or until meat is fork-tender. Add more water if needed during cooking time.

Tomato Swiss Steak

4 servings

¼ cup all-purpose flour
1 teaspoon salt
¼ teaspoon pepper
1½ to 2 pounds round steak

2 large onions, sliced
1 can (10½ ounces) condensed tomato soup
1 can (8 ounces) tomato sauce

1. Combine flour, salt, and pepper. Place steak on a board and pound half of the flour mixture into each side of the steak with the back of a heavy knife. Cut meat in 4 pieces and place in an 8-inch square glass baking dish. Sprinkle any remaining flour mixture over meat. Spread onions over meat. Combine tomato soup, tomato sauce, and 1 soup can of water. Pour over steak.
2. Cook, covered, on HIGH for 5 minutes. Cook on BAKE for 20 minutes. Turn meat over. Cook, covered, on BAKE for 20 minutes, or until meat is tender. Check meat during last half of cooking time and add more water if needed.

Beef Goulash

4 servings

2 pounds stew beef, cut in 1-inch cubes
3 to 4 large tomatoes
1 onion, coarsely chopped

1 teaspoon salt
½ teaspoon freshly ground pepper
1 cup sour cream (optional)

1. Place beef in a 2- to 3-quart casserole.
2. Peel tomatoes; remove cores. Cut tomatoes in chunks. Place in casserole with beef, onion, salt, and pepper. Toss mixture lightly.
3. Cook, covered, on REHEAT for 40 to 45 minutes, or until beef is tender. Stir occasionally during cooking period.
4. If desired, stir sour cream into mixture and let stand, covered, 5 minutes.

Note: This is excellent served with cooked egg noodles.

Beef Casserole

4 servings

1 package (10 ounces) frozen French-style green beans
1 pound stew beef, cut in cubes
½ teaspoon meat tenderizer
1 large onion, chopped

1 can (10½ ounces) condensed tomato soup, undiluted
2 cups cooked egg noodles
Salt and pepper

1. Place green beans in a 1-quart casserole, icy side up.
2. Cook, covered, on HIGH for 5 minutes. Reserve.
3. Combine beef cubes and meat tenderizer in a 2- to 3-quart casserole. Toss lightly. Add onion.
4. Cook, covered, on HIGH for 15 to 20 minutes, or until meat is tender.
5. Add tomato soup, ½ cup water, noodles, and green beans to beef cubes. Stir lightly and season with salt and pepper to taste.
6. Cook, covered, on HIGH for 5 minutes.
7. Stir and let stand, covered, 3 to 4 minutes before serving.

Short Ribs of Beef

4 servings

2 pounds meaty short ribs of beef
1 clove garlic, minced
½ teaspoon salt

½ cup dry red wine
1 tablespoon liquid gravy seasoning

1. Arrange short ribs in a 2- or 3-quart casserole. Sprinkle with garlic and salt. Combine wine and liquid gravy seasoning. Pour over short ribs.
2. Cook, covered, on ROAST for about 30 minutes, or until meat is tender, stirring once or twice during cooking period.
3. Remove and let stand 5 minutes.

Oriental Beef

4 to 6 servings

1½ to 2 pounds thinly cut boneless
   sirloin steak, cut into thin strips
½ cup soy sauce
¼ cup dry sherry
¼ cup water
1 tablespoon sugar

1 whole clove garlic
2 thin slices ginger
1 bunch green onions
1 can (5 ounces) water chestnuts
½ bunch fresh broccoli
½ pound fresh bean sprouts

1. Combine steak, soy sauce, sherry, water, sugar, garlic, and ginger in a 12- by 7-inch glass baking dish. Cover with plastic wrap and let stand at room temperature 4 hours. Occasionally, uncover and turn meat over.
2. Clean green onions and cut into 2-inch pieces. Drain water chestnuts and slice. Peel broccoli and cut stems into very thin slices. Leave flowerets whole. Rinse bean sprouts.
3. Remove garlic and ginger from meat. Place some of each vegetable in each corner of the baking dish. Cover dish with plastic wrap.
4. Cook on HIGH for 10 to 12 minutes. Remove and let stand 2 minutes.
5. Serve with hot cooked rice.

Beef and Peppers

4 servings

2 tablespoons cooking oil
1 pound top round or sirloin steak,
   cut into thin strips
1 medium onion, finely chopped
1 clove garlic, minced
1 teaspoon salt

⅛ teaspoon pepper
1 can (1 pound) tomatoes, broken up
2 large green peppers, seeded and
   cut in strips
2 tablespoons soy sauce

1. Put oil in a 2- or 3-quart casserole or baking dish. Add beef strips and toss lightly so that meat is coated with oil. Add onion, garlic, salt, and pepper.
2. Cook, covered, on HIGH for about 5 minutes, stirring once during cooking time.
3. Add tomatoes. Cook, covered, on HIGH for 4 minutes, stirring once during cooking time.
4. Add green pepper strips and soy sauce and toss thoroughly.
5. Cook, covered, on HIGH for 4 to 5 minutes, or just until green pepper is tender.
6. Serve with hot cooked rice or chow mein noodles.

Broccoli and Beef

2 servings

1 large or 2 medium stalks broccoli
2 tablespoons cooking oil
½ cup coarsely chopped onions
1 clove garlic, minced

½ pound ground beef
½ cup thinly sliced celery
2 teaspoons soy sauce
1 tablespoon dry sherry

1. Cut off flowerets from top of broccoli and cut large flowers in halves or quarters. Peel broccoli stalks and cut in diagonal slices about ½ inch thick. Set aside.

30

2. In a 1½-quart casserole place oil, onion, and garlic. Add beef, broken into pieces.
3. Cook, covered, on HIGH for 4 minutes.
4. Remove casserole and break up pieces of beef with a fork. Add celery, soy sauce, sherry, and broccoli. Stir.
5. Cook, covered, on REHEAT for 8 to 9 minutes, or just until broccoli is crisply tender.
6. Serve over hot cooked rice with additional soy sauce.

Shish Kabob   6 to 8 servings

½ cup wine vinegar
½ cup cooking oil
1 teaspoon onion salt
1 clove garlic, split in half
¼ cup soy sauce
2 teaspoons Italian seasoning
2 pounds boneless sirloin or top round steak

½ pound small fresh mushrooms
1 dozen tomato wedges or cherry tomatoes
1 green pepper, seeded and cut in 1-inch squares

1. Combine vinegar, oil, onion salt, garlic, soy sauce, Italian seasoning, and ½ cup water in a large mixing bowl. Cut steak in 1-inch cubes. Add to marinade and let stand at room temperature 5 to 6 hours.
2. Place meat cubes and desired vegetables alternately on long wooden skewers.
3. Place 2 or 3 skewers on a dinner plate.
4. Cook, uncovered, on HIGH for 3 minutes for medium rare. Cook slightly longer for well-done meat.

Note: The marinade turns the meat brown while standing, for attractive color in the finished kabobs.

Beef Rolls   6 servings

2 pounds beef top round steak, cut ½ inch thick
2 tablespoons butter or margarine
½ cup chopped celery with leaves
¼ cup chopped onion
1 cup soft bread crumbs

¼ teaspoon crushed rosemary
¼ teaspoon thyme
Grind of fresh pepper
1 can (10½ ounces) condensed cream of mushroom soup

1. Pound steak lightly with the side of a cleaver or a meat pounder. Cut into 6 pieces.
2. Combine butter, celery, and onion in a 1-quart glass measure. Cook on HIGH for 4 minutes, or until onion is limp. Remove from oven. Stir in bread crumbs, rosemary, thyme, and pepper.
3. Divide stuffing among pieces of steak. Roll meat around stuffing and fasten with toothpicks. Arrange in a glass baking dish. Spoon soup over top of meat rolls.
4. Cook on ROAST for 20 minutes, or until meat is tender. Rearrange meat once during cooking time, spooning sauce over top of meat rolls.
5. Let stand 5 minutes before serving.

Beef Stew

4 to 6 servings

- 2 pounds stew meat, cut in 1-inch cubes
- ½ teaspoon meat tenderizer
- ½ teaspoon salt
- 1 package (1½ ounces) brown gravy mix with mushrooms
- 3 stalks celery, cut in chunks
- 3 medium carrots, peeled and sliced
- 3 medium potatoes, peeled and cut in eighths

1. Put beef cubes in a 3-quart casserole or baking dish. Sprinkle with meat tenderizer and salt. Combine gravy mix with 1 cup water and stir well. Pour over meat.
2. Cook, covered, on HIGH for 5 minutes. Cook, covered, on BAKE for 20 minutes.
3. Add celery, carrots, and potatoes, and stir lightly so that vegetables are covered with gravy. Add more water for a thinner gravy and a little additional salt for the vegetables, if desired.
4. Cook, covered, on SIMMER for 25 to 30 minutes, or until vegetables and meat are tender.
5. Remove and let stand 5 minutes.

Beef Stroganoff

2 to 3 servings

- 1 pound beef sirloin steak, cut in very thin strips
- 1 large onion, chopped
- 1 bouillon cube
- 1 tablespoon all-purpose flour
- 1 tablespoon catsup
- ½ teaspoon salt
- ½ cup water
- ½ cup sour cream
- Hot cooked rice or noodles

1. Combine beef strips, onion, bouillon cube, flour, catsup, salt, and water in a 1-quart casserole or baking dish. Stir well.
2. Cook, covered, on HIGH for 4 minutes. Stir well. Cook, covered, on HIGH for 4 minutes, or until meat is tender.
3. Stir in sour cream. Cook on WARM for 3 minutes.
4. Let stand 3 minutes.
5. Serve with hot cooked rice or noodles.

Corned Beef

about 6 servings

- 3-pound corned beef brisket
- 1 bay leaf
- ½ teaspoon whole black peppercorns
- 2 cloves garlic
- 4 whole cloves
- 2 cups water
- 2 medium onions, quartered
- 3 carrots, peeled and sliced
- 3 potatoes, peeled and quartered
- 1 very small cabbage, cut into 8 wedges

1. Place corned beef in a 3- to 4-quart casserole or baking dish. Add bay leaf, peppercorns, garlic, cloves, water, and onions. Cook, covered, on HIGH for 15 minutes.
2. Turn meat over. Cook on SIMMER for 30 minutes. Turn meat over and cook on SIMMER for 30 minutes, or until meat is almost tender.
3. Add carrots, potatoes, and cabbage wedges. Cook, covered, on SIMMER for 30 minutes, or until meat and vegetables are tender.
4. Let stand 5 minutes before slicing meat.

Herbed Meat Loaf
6 servings

2 eggs, lightly beaten
¼ cup bread crumbs
1 tablespoon chopped parsley
¼ cup chopped chives or onion
1 tablespoon dried basil
¼ cup chopped green pepper

1½ teaspoons salt
½ teaspoon pepper
1 pound ground beef
½ pound ground veal
½ pound lean ground pork

1. Combine all ingredients except meat, and mix well. Add ground meat and blend with hands, being careful not to mix more than necessary.
2. Shape into a round and place in an 8-inch pie plate. Make a depression in center of loaf.
3. Cook, uncovered, on ROAST for 25 minutes, or until center of loaf is cooked.
4. Remove. Cover meat and let stand 10 minutes before serving.
5. Serve with mushroom gravy or tomato sauce, if desired.

Note: If using probe, insert in center of meat loaf. Set Cook Control for 155°F. and START. Let stand, covered, 5 to 10 minutes before serving.

Our Favorite Meat Loaf
6 servings

1 can (8 ounces) tomato sauce
¼ cup brown sugar
¼ cup vinegar
1 teaspoon prepared mustard
1 egg, lightly beaten

1 medium onion, minced
¼ cup cracker crumbs
2 pounds ground beef
1½ teaspoons salt
¼ teaspoon pepper

1. Combine tomato sauce, brown sugar, vinegar, and mustard in a small bowl. Set aside.
2. Combine egg, onion, cracker crumbs, ground beef, salt, and pepper in a mixing bowl. Add ½ cup of the tomato mixture and blend thoroughly. Shape into an oval loaf and place in an oblong baking dish. Make a depression in top of loaf. Pour remaining tomato sauce over top of meat.
3. Cook, uncovered, on ROAST for 25 to 30 minutes, or until center is cooked.
4. Cover meat and let stand about 10 minutes before serving.

Note: If using probe, insert in center of meat loaf. Set Cook Control for 155°F. and START. Let stand, covered, 5 to 10 minutes before serving.

Vegetable Meat Loaf
6 servings

1 can (10½ ounces) condensed
    vegetable soup, undiluted
2 pounds lean ground beef
½ cup fine dry bread crumbs
½ cup chopped onions
2 tablespoons chopped parsley

1 tablespoon Worcestershire sauce
1 egg, lightly beaten
1 teaspoon salt
Freshly ground pepper to taste
4 tomato slices
½ cup grated process American cheese

1. Combine soup, beef, bread crumbs, onion, parsley, Worcestershire, egg, salt, and pepper. Mix thoroughly. Shape into an oval loaf and place in an oblong baking dish. Make a slight depression in the top of the loaf.

2. Cook, covered with a piece of waxed paper, on ROAST for 25 to 30 minutes, or until center of loaf is done.
3. Place tomato slices and cheese on top of loaf. Cook, uncovered, on BRAISE for 5 minutes, or until cheese is melted.
4. Let stand at WARM for 5 minutes before serving.

Note: If using probe, insert in center of meat loaf. Set Cook Control for 155°F. and START. Let stand, covered, 5 to 10 minutes before serving.

Limas and Beef Casserole

4 to 6 servings

1 pound lean ground beef
1 clove garlic, crushed
1 medium onion, chopped
1 small green pepper, seeded and chopped
¼ teaspoon chili powder

½ teaspoon dry mustard
2 teaspoons Worcestershire sauce
½ teaspoon salt
2 cans (1 pound each) lima beans
1 can (8 ounces) tomato sauce

1. Combine beef, garlic, onion, and green pepper in a 2- or 3-quart casserole.
2. Cook, uncovered, on HIGH for 5 minutes. Stir with a fork to break up meat.
3. Add remaining ingredients. Toss lightly.
4. Cook, covered, on HIGH for 6 minutes, stirring once during cooking period.
5. Let stand, covered, 3 to 4 minutes before serving.

Cabbage Rolls

6 to 8 servings

1 head cabbage, about 1½ pounds
1 pound ground beef
½ pound ground pork
¾ cup cooked rice
1 egg, lightly beaten
1 teaspoon thyme

1 tablespoon chopped parsley
1 clove garlic, minced
2 teaspoons salt
¼ teaspoon pepper
¼ cup butter or margarine
1 can (16 ounces) tomato sauce

1. Remove core from cabbage. Remove any blemished leaves. Put cabbage in a 3-quart casserole. Add boiling water to cover about one-quarter of the bottom of the cabbage.
2. Cook, covered, on HIGH for 6 minutes.
3. Cool cabbage slightly and remove 6 to 8 of the large outside leaves. Remove the tough center core from each leaf.
4. Combine ground beef, pork, rice, egg, thyme, parsley, garlic, salt, and pepper. Toss together lightly. Divide mixture among cabbage leaves. Wrap leaves tightly around mixture.
5. Line bottom of a 3-quart casserole with some of the leftover cabbage leaves. Place rolled cabbage packets on top of loose leaves. Cover with remaining pieces of cabbage.
6. Top with butter. Pour tomato sauce over top.
7. Cook, covered, on REHEAT for about 30 minutes, or until meat is cooked and rolls are fork-tender.
8. Remove and let stand, covered, 10 minutes.
9. Discard the top loose cabbage leaves before serving.

Beef Tacos

10 to 12 servings

1 pound ground beef
½ cup chopped onions
1 can (8 ounces) tomato sauce
¼ teaspoon chili powder
¼ teaspoon salt
¼ teaspoon garlic salt

10 to 12 fully cooked taco shells
Grated Cheddar cheese
Shredded lettuce
Chopped fresh tomato
Finely diced avocado

1. Crumble beef into a 1-quart casserole. Add onion.
2. Cook, uncovered, on HIGH for 4 minutes. Stir with a fork to break up meat.
3. Add tomato sauce, chili powder, salt, and garlic salt.
4. Cover casserole with a paper towel. Cook for about 5 minutes, or until sauce is blended and thickened.
5. Cook, covered, on BAKE for 6 to 7 minutes.
6. Spoon filling into taco shells. Serve immediately with cheese, lettuce, tomato, and avocado in side dishes to sprinkle over top of hot filling.

Meatballs Stroganoff

5 to 6 servings

½ cup milk
3 slices bread, cubed
1 pound ground beef
1 egg, lightly beaten
3 tablespoons grated onion
3 tablespoons dried parsley flakes
1 teaspoon salt

Freshly ground pepper to taste
3 tablespoons flour
1½ cups beef broth or bouillon
1 tablespoon tomato paste
¼ teaspoon paprika
1½ cups dairy sour cream

1. Pour milk into a mixing bowl. Cook in oven on ROAST for about 1 minute, or just long enough to warm milk.
2. Add bread cubes and mix until all the milk is absorbed by the bread.
3. Add beef, egg, onion, parsley, salt, and pepper. Blend well. Form mixture into balls about 1½ inches in diameter. Arrange meatballs in a 7¼- by 11¾-inch oblong baking dish.
4. Cook, covered, on HIGH for 3 minutes.
5. Turn meatballs over. Cook, covered, on HIGH for 2 minutes. Drain off fat and liquid.
6. Stir together flour and beef broth to make a smooth mixture. Stir in tomato paste and paprika. Pour mixture over top of meatballs.
7. Cook, covered, on BAKE for 5 minutes, stirring occasionally.
8. Top with sour cream and stir lightly. Cover and cook on LOW for 5 minutes, or until serving time.
9. Serve with cooked rice or hot buttered noddles.

Onion Meatballs

6 servings

1½ pounds lean ground beef	3 tablespoons all-purpose flour
½ cup milk	2 tablespoons chopped parsley
1 package (1¼ ounces) onion soup mix	½ cup dairy sour cream

1. In a large bowl combine beef with milk and 2 tablespoons of the onion soup mix. Mix thoroughly. Shape into 24 meatballs. Place in a 3-quart oblong baking dish.
2. Cook, covered, on HIGH for 3 minutes.
3. Turn meatballs over. Cook, covered, on HIGH for 2 minutes.
4. Remove meatballs. Stir flour into drippings in dish. Stir in 1¹/₂ cups water, parsley, and remaining soup mix.
5. Cook, uncovered, on BAKE for about 5 minutes, or until mixture comes to a boil.
6. Add meatballs. Cook, covered, on BAKE for 6 minutes, stirring occasionally during cooking time.
7. Gradually blend in sour cream. Cover and let stand 5 minutes before serving.
8. Serve over rice or noodles.

Nutty Meatballs

4 servings

1 pound lean ground beef	2 tablespoons chopped parsley
1 egg, lightly beaten	½ teaspoon salt
¼ cup milk	Freshly ground pepper to taste
½ cup soft bread crumbs	1 can (10½ ounces) condensed tomato soup, undiluted
½ cup chopped pecans	

1. Combine ground beef with egg, milk, bread crumbs, pecans, parsley, salt, and pepper. Blend well. Shape mixture into 24 meatballs. Arrange in an oblong baking dish.
2. Cook, covered, on ROAST for 5 minutes.
3. Turn meatballs over. Cook, covered, on ROAST for 2 minutes.
4. Drain off fat and liquid. Add tomato soup. Blend lightly. Cook, covered, on ROAST for 3 minutes, or until piping hot, stirring once during cooking time.
5. Let stand 5 minutes before serving.
6. Serve over hot mashed potatoes.

Hamburger Hash Burgundy

4 servings

1 pound ground beef	½ cup dry red wine
½ cup chopped onions	2 cups diced raw potatoes
2 tablespoons all-purpose flour	½ cup diced celery
1 can (10½ ounces) condensed beef consomme, undiluted	Salt and pepper to taste

1. Crumble ground beef into a 2-quart casserole. Stir in onion.
2. Cook, uncovered, on HIGH for 2 minutes.
3. Stir with a fork to break up meat. Cook on HIGH for 4 minutes, or until meat loses its red color.

4. Sprinkle flour on meat. Stir in well. Add consomme and wine.
5. Cook, covered, on HIGH for 5 minutes.
6. Add potatoes and celery, and salt and pepper to taste.
7. Cook, covered, on BAKE for 15 to 20 minutes, or until potatoes are tender.
8. Let stand 5 minutes before serving.

Burger Stroganoff

4 to 6 servings

1 pound ground beef
½ cup minced onions
1 clove garlic, minced
1 pound fresh mushrooms, sliced

2 teaspoons salt
¼ teaspoon pepper
2 tablespoons all-purpose flour
1 cup dairy sour cream

1. Crumble beef into a 1½-quart casserole. Add onion and garlic.
2. Cook, uncovered, on HIGH for 2 minutes. Stir lightly.
3. Cook, uncovered, on HIGH for 2 minutes.
4. Add mushrooms. Cook, covered, on HIGH for 2 minutes.
5. Sprinkle salt, pepper, and flour over top of meat. Stir well.
6. Cook, covered, on HIGH for 3 minutes, stirring once during cooking period.
7. Stir in sour cream. Cook, covered, on WARM for 3 minutes, or just long enough to warm the sour cream mixture. It can be left in oven on WARM for several minutes before serving.
8. Serve over hot cooked noodles or mashed potatoes.

Chili Con Carne

4 servings

1 pound lean ground beef
½ cup chopped onion
¼ cup chopped green pepper
1 clove garlic, minced

2 to 3 teaspoons chili powder
1 teaspoon salt
1 can (16 ounces) whole tomatoes
1 can (1 pound) kidney beans

1. Crumble ground beef into a 2-quart casserole or baking dish. Add onion, green pepper, and garlic.
2. Cook, uncovered, on HIGH for 4 minutes. Break up meat with a fork.
3. Add remaining ingredients. Cook, covered, on ROAST for 10 to 12 minutes, or until piping hot. Stir once during cooking period.
4. Let stand 5 minutes before serving.

Veal Elegante

3 to 4 servings

1 pound boneless veal, cut in cubes
½ cup minced onions
1 fresh tomato, peeled and cubed

½ cup dry white wine
Salt and pepper to taste

1. Combine veal, onion, tomato, wine, and salt and pepper to taste in a 2-quart casserole.
2. Cook, covered, on HIGH for 15 minutes, or until veal is tender. Stir once during cooking period.
3. Let stand 3 to 4 minutes. Serve over hot cooked rice or cooked buttered noodles.

Veal Parmigiana

4 servings

1 egg, lightly beaten
¼ teaspoon salt
3 tablespoons cracker crumbs
⅓ cup grated Parmesan cheese
1 pound veal cutlets
2 tablespoons cooking oil
¼ cup dry vermouth

1 medium onion, chopped
1 cup (4 ounces) sliced or shredded
  mozzarella cheese
1 can (8 ounces) tomato sauce
Grind of fresh pepper
⅛ teaspoon oregano

1. Beat egg with salt in a shallow dish. Combine cracker crumbs and Parmesan cheese on a piece of waxed paper. Cut veal into 4 serving pieces. Place each piece between two pieces of waxed paper and pound to about ¼ inch thick with the side of a cleaver.
2. Dip veal in egg and then in cracker crumbs. Heat oil in a skillet on top of the range. Cook veal in hot oil until golden brown on both sides.
3. Place veal in a 10- by 6-inch baking dish. Add vermouth to skillet and heat about 1 minute, scraping up browned bits from bottom of skillet. Pour over veal cutlets.
4. Sprinkle onion over meat. Top with mozzarella cheese. Spoon tomato sauce over top, and season with pepper and oregano.
5. Cook, covered, on BAKE for 10 minutes, or until sauce is bubbly and cheese is melted.

Cordon Bleu Veal

2 servings

½ pound veal round steak or cutlets,
  cut ½-inch thick
1 slice Swiss cheese
2 thin slices boiled ham
1½ tablespoons all-purpose flour

1 egg
¼ cup dry bread crumbs
1½ tablespoons butter or margarine
1 tablespoon chopped parsley
1 tablespoon dry white wine

1. Cut veal into 4 pieces. Place each piece of veal between 2 sheets of waxed paper and pound with the side of a cleaver or a mallet until veal is ⅛ inch thick.
2. Cut cheese in 2 pieces. Fold each piece of cheese in half. Place on a slice of ham. Roll ham around cheese three times so that finished roll of ham is smaller than the pieces of veal. Place ham on one slice of veal and top with a second slice. Press edges of veal together to seal.
3. Put flour on a piece of waxed paper. Beat egg lightly with 1 tablespoon water. Put bread crumbs on a piece of waxed paper. Dip veal in flour, then in beaten egg, and finally coat well with bread crumbs.
4. Put butter and parsley in an 8-inch baking dish.
5. Cook in oven for 1 minute, or just long enough to heat butter well.
6. Add veal to very hot butter.
7. Cook, uncovered, on HIGH for 2 minutes. Turn veal slices over.
8. Cook, uncovered, on HIGH for 2 minutes, or just until veal is tender.
9. Pour wine over veal and tilt pan to swish wine around. Serve immediately.

Zesty Lamb Chops

4 servings

4 shoulder lamb chops
½ cup coarsely chopped onion
1 clove garlic, minced
½ cup catsup
2 tablespoons Worcestershire sauce
1 tablespoon prepared mustard

1. In a baking dish arrange lamb chops in one layer. Sprinkle with onion and garlic. Cook, uncovered, on HIGH for 5 minutes.
2. Combine remaining ingredients and spread over lamb chops. Cook, covered, on BAKE for 15 minutes, or until lamb chops are tender.

Lamb Stew

4 servings

1 pound boneless lamb, cut in 1-inch cubes
1 package (⅝ ounce) brown gravy mix
2 tablespoons all-purpose flour
1 teaspoon salt
⅛ teaspoon pepper
1 clove garlic, minced
½ teaspoon Worcestershire sauce
¼ cup red wine
3 medium carrots, peeled and cut in chunks
2 stalks celery, cut in pieces
2 potatoes, peeled and cut in cubes

1. In a 2- or 3-quart casserole combine lamb and gravy mix.
2. Cook, uncovered, on HIGH for 5 minutes, stirring occasionally.
3. Add remaining ingredients, with 1 cup of water. Stir well.
4. Cook, covered, on BAKE for 20 to 25 minutes, or until meat and vegetables are tender. Stir once during cooking period.
5. Let stand 3 to 4 minutes before serving.

Curried Lamb

4 servings

1 pound boneless lamb, cut in 1-inch cubes
2 tablespoons all-purpose flour
1 clove garlic, minced
1 large onion, sliced
¼ cup butter or margarine
1½ tablespoons curry powder
2 apples, peeled, cored, and chopped in coarse pieces
2 tablespoons seedless raisins
1½ teaspoons salt

1. Toss lamb cubes lightly with flour. Set aside.
2. Place garlic, onion, butter, and curry powder in a 2-quart casserole. Cook, uncovered, on HIGH for 3 to 4 minutes.
3. Add lamb and toss lightly.
4. Cook, uncovered, on HIGH for 3 minutes, stirring once.
5. Add apples, raisins, and salt. Stir in ½ cup water.
6. Cook, covered, on BAKE for 20 to 25 minutes, or until meat is tender. Stir once during cooking period.
7. Cook, covered, on WARM for 5 minutes.
8. Serve with hot cooked rice and accompaniments, such as flaked coconut, chopped peanuts or walnuts, and chutney.

Leg of Lamb

8 servings

4- to 4½-pound leg of lamb, bone in   **1 large clove garlic, cut in thin slices**

1. Cut small slits in both sides of leg of lamb. Insert thin slices of garlic in slits.
2. Place leg of lamb, fat side down, on a microwave roasting rack in an oblong baking dish. Use an inverted saucer if there is no rack.
3. Cook, uncovered, on HIGH for 25 minutes.
4. Turn lamb over with fat side up. Insert a microwave meat thermometer in thickest part of leg, making sure not to touch the bone. Cook, uncovered, on ROAST for about 20 minutes, or until thermometer registers 150°F. Remove from oven, insert a standard meat thermometer into meat to register 150°F.
5. Cover with aluminum foil and let stand, about 20 minutes, until thermometer reaches desired temperature and to make the meat easier to carve.

Note: Cooking time for leg of lamb averages 9 to 11 minutes per pound.

Loin of Pork

6 to 8 servings

1 loin of pork, about 4 pounds

1. Place roast, fat side down, on a roasting rack or inverted saucer in a 12- by 7-inch glass baking dish. Cook, uncovered, on HIGH for 10 minutes.
2. Turn loin of pork fat side up. Insert probe in thickest part of meat, making sure that it does not touch bone or fat pocket.
3. Cover lightly with a piece of waxed paper. Set Temperature Control for 160°F.
4. Let stand 10 minutes before serving.

Note: If you do not use the probe, roasting time is about 10 to 12 minutes per pound with the oven set on BAKE.

Orange Ginger Pork Chops

6 servings

6 lean pork chops   **1 teaspoon ground ginger**
¼ cup orange juice   **1 large orange, peeled and sliced**
½ teaspoon salt   **Dairy sour cream**

1. Trim fat from pork chops. Place in an oblong baking dish. Pour orange juice over chops.
2. Cook, covered, on HIGH for 10 minutes.
3. Remove from oven. Turn chops over. Sprinkle with salt and ginger. Place an orange slice on top of each chop.
4. Cook, covered, on HIGH for 10 minutes.
5. Remove from oven. Top each chop with a dollop of sour cream. Cover and let stand 5 minutes before serving.

Following pages; Left, *Stuffed Pork Chops (recipe, p. 46);* right, *Barbecued Spareribs (recipe, p. 47)*

Stuffed Pork Chops *(Illustrated on page 44)* 4 servings

1 cup coarse dry bread crumbs	Freshly ground pepper to taste
¾ cup chopped apples	Pinch of sage
3 tablespoons chopped raisins	2 tablespoons melted butter or
½ teaspoon salt	margarine
2 tablespoons sugar	8 thin rib or loin pork chops
2 tablespoons finely minced onion	½ package (⅝ ounce) brown gravy mix

1. Combine bread crumbs, apples, raisins, salt, sugar, onion, pepper, sage, and melted butter. Toss together lightly. Moisten slightly with hot water if stuffing is dry.
2. Trim all fat from pork chops. Place 4 chops in bottom of an 8-inch square baking dish. Divide stuffing into 4 portions and place one portion on top of each chop. Cover chops with 4 remaining chops, pressing together lightly.
3. Sprinkle brown gravy mix over top of chops. (To make an even layer, sift mixture through a small strainer.)
4. Cook, uncovered, on HIGH for 15 minutes, or just until pork is tender. Do not overcook.

Pork Chops with Apricots 4 servings

4 center-cut pork chops, about 1½ pounds	½ teaspoon oregano
	Salt and pepper to taste
2 tablespoons brown sugar	1 can (8¾ ounces) apricot halves

1. Trim all fat from pork chops. Place chops in an 8-inch square baking dish. Sprinkle brown sugar, oregano, salt, and pepper on each chop. Pour half of the liquid from the apricots over chops.
2. Cook, covered, on HIGH for 10 minutes.
3. Spoon some of liquid over top of chops. Top with apricots.
4. Cook, covered, on HIGH for 10 minutes, or until pork is tender.
5. Let stand 5 minutes before serving.

Tomatoed Pork Chops 4 servings

4 center-cut pork chops	Dash of hot-pepper sauce
1 large onion, sliced	¼ teaspoon marjoram
1 can (15 ounces) tomato sauce	

1. Trim all fat from chops. Place chops in an 8-inch square baking dish. Cover with sliced onion. Combine tomato sauce, hot-pepper sauce, and marjoram. Pour mixture over chops.
2. Cook, covered, on HIGH for 10 minutes.
3. Remove from oven. Turn chops over. Spoon some of the sauce in bottom of dish over chops.
4. Cook, covered, on HIGH for 10 minutes.
5. Allow to stand 5 minutes before serving.

Barbecued Spareribs *(Illustrated on page 45)*   4 servings

2 tablespoons cooking oil	2 teaspoons Worcestershire sauce
½ cup minced onions	1 teaspoon prepared mustard
2 cans (8 ounces each) tomato sauce	1 teaspoon salt
2 tablespoons lemon juice	¼ teaspoon black pepper
2 tablespoons brown sugar	1½ pounds spareribs
1 tablespoon white sugar	

1. Put oil and onion in a 1-quart casserole. Cook, covered, on HIGH for 3 to 4 minutes.
2. Add 2 tablespoons of water and remaining ingredients except spareribs. Cover and cook on HIGH for 3 minutes.
3. Cover and let stand.
4. Cut ribs apart between bones and place in a 3-quart casserole.
5. Cook, covered, on ROAST for 15 minutes.
6. Pour off fat and juices. Cover ribs with ¾ cup of the barbecue sauce mixture.
7. Cook, uncovered, on ROAST for 15 minutes.
8. Turn ribs over. Spoon ½ cup sauce over top. Cook on ROAST for 10 to 20 minutes, or until meat is tender.
9. Let stand 5 minutes before serving.

Note: This recipe makes about 2½ cups sauce. Save remainder for other ribs or use on chicken or barbecued steak.

Sweet and Sour Pork   4 to 6 servings

1½ pounds lean pork shoulder, cut into ½-inch cubes	½ cup vinegar
1 small onion, sliced	½ cup sugar
1 teaspoon salt	1 can (8 ounces) pineapple chunks, drained
1 cup chicken bouillon	2 large green peppers, seeded and cut in squares
2 teaspoons soy sauce	Hot cooked rice
2½ tablespoons cornstarch	

1. Put pork cubes, onion, and salt in a 3-quart casserole or baking dish.
2. Cook, covered, on HIGH for 18 minutes. Stir once.
3. Drain off pork fat.
4. Combine bouillon, soy sauce, cornstarch, vinegar, and sugar. Pour over pork.
5. Cook, covered, on ROAST for 25 minutes, or until pork is tender. Stir once during cooking time.
6. Add pineapple and green peppers. Cook on HIGH for 4 minutes, stirring once during cooking time.
7. Let stand 5 minutes before serving.
8. Serve with hot cooked rice.

Baked Ham

10 to 12 servings

½ precooked smoked ham,
   about 4 pounds
1 can (4 ounces) pineapple,
   sliced, drained, juice
   reserved

¼ cup brown sugar
Whole cloves

1. Place ham, fat side down, on a roasting rack or inverted saucer in a 12- by 7-inch glass baking dish. Cook on ROAST for 10 minutes.
2. Remove ham from oven. Turn fat side up. Combine 2 tablespoons of the pineapple juice with the brown sugar to make a paste. Spread over top of ham. Put pineapple slices on top and stud with cloves. Insert probe into thickest part of ham, making sure probe does not touch the bone or a fat pocket.
3. Set Cook Control for 120°F. Set Temperature Control on ROAST.
4. Let stand, covered with foil, about 10 minutes before carving.

Baked Ham Steak

2 servings

1 precooked ham steak, about 1
   pound

2 teaspoons brown sugar
½ teaspoon prepared mustard

1. Put ham steak in an 8-inch square baking dish.
2. Cook, covered, on ROAST for 4 minutes.
3. Remove from oven and drain off juices. Mix juices with brown sugar and mustard. Spread over top of steak.
4. Cook, uncovered, on ROAST for 2 minutes.
5. Cover loosely and let stand 2 or 3 minutes before serving.

Scalloped Potatoes and Ham

4 servings

1½ cups milk
2 cups leftover pieces of baked ham
4 cups sliced potatoes
⅔ cup chopped onions
2 tablespoons all-purpose flour

1 teaspoon salt
⅛ teaspoon pepper
2 tablespoons butter or margarine
Paprika

1. Put milk in a 2-cup measure.
2. Cook, covered with waxed paper, on HIGH for 2 to 3 minutes, or just long enough to warm milk.
3. Put a layer of ham in bottom of a 3-quart glass casserole. Add a layer of potatoes and onions. Sprinkle with flour, salt, and pepper. Dot with butter. Add another layer of ham, potatoes, and onions. Pour warm milk over layers. Sprinkle with paprika.
4. Cook, covered, on HIGH for 10 minutes.
5. Cook, uncovered, on ROAST for 15 minutes, or until potatoes are tender.
6. Cover loosely with plastic wrap and let stand 2 minutes before serving.

Ham Casserole

4 servings

1½ cups diced cooked ham
2 tablespoons chopped onion
⅛ teaspoon tarragon
2 tablespoons butter or margarine
1 can (10½ ounces) condensed cream
  of chicken soup, undiluted

1½ cups cooked narrow egg noodles
½ cup cooked French-style green
  beans
2 tablespoons buttered bread crumbs

1. Combine ham, onion, tarragon, and butter in a 1½-quart casserole.
2. Cook, covered, on HIGH for 3 minutes.
3. Add chicken soup, egg noodles, green beans, and ½ cup water. Toss lightly.
4. Cook, covered, on HIGH for 5 minutes, or until piping hot. Stir once during cooking period.
5. Top with crumbs. Cover and let stand 2 to 3 minutes.

Bacon

8 strips

8 slices bacon

1. Separate bacon strips and lay 4 slices in an oblong baking dish on 2 layers of paper towels. Cover with 1 piece of paper towel. Place 4 more slices bacon on towel. Cover with towel.
2. Cook on HIGH for 5, to 7 minutes, or to the desired degree of crispness.

Note: Cooking bacon in the microwave oven has more variables than almost any other food. Cooking time depends on the thickness of the slice, the amount of fat in the bacon, and the desired degree of crispness. It may be advisable to cook it the first time at the times suggested here, and thereafter cook it according to your own findings.

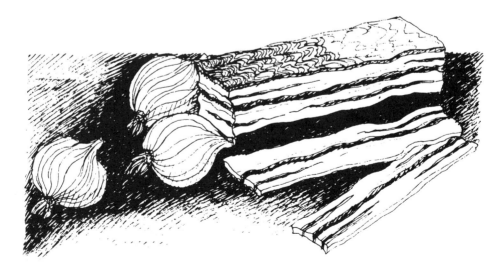

Poultry

Frozen whole chickens and turkeys can be successfully thawed in the microwave oven. Cover spots that begin to cook with bits of aluminum foil, so that when the bird is cooked you will not have overdone spots. Poultry should be thawed completely before cooking.

When roasting a bird, check for doneness by removing the bird from the oven and inserting the thermometer. When the temperature is fifteen degrees below the final temperature desired, it is time to let the bird stand—it will finish cooking out of the oven. Stuffing will not increase the cooking time. Skin may be crisped under a conventional oven broiler.

When cooking chicken parts, put the larger pieces of chicken on the outer edges of the dish and the smaller pieces in the center, so that all pieces will cook evenly.

Poultry—How to Defrost

1. Defrost in original wrapping. Remove any metal twist-ties or clamps.
2. Place poultry in glass baking dish.
3. Poultry is thawed on the DEFROST setting, unless otherwise stated.
4. Poultry will be cool in the center when removed from the oven. Wing and leg tips may begin cooking before center is thoroughly defrosted.

Poultry	Minutes per Pound	Setting	Standing Time	Special Notes
Turkey:				
8 pounds and under	3 to 5	DEFROST	1 hour	Turn over once. Immerse in cold water during standing time.
over 8 pounds	3 to 5	ROAST	1 hour	
Turkey breast:				
4 pounds and under	3 to 5	DEFROST	20 minutes	
over 4 pounds	1	ROAST		Begin on ROAST; turn over;
	2	SIMMER	20 minutes	continue on SIMMER.
Turkey drumsticks	5 to 6	DEFROST	15 to 20 minutes	Turn every 5 minutes. Separate pieces when partially thawed.
Chicken (whole)	6 to 8	DEFROST	20 to 25 minutes	Turn over once. Immerse in cold water during standing time.
Chicken (cut-up)	5 to 6	DEFROST	10 to 15 minutes	Turn every 5 minutes. Separate pieces when partially thawed.
Duckling	4	ROAST	20 to 30 minutes	Turn over once. Immerse in cold water during standing time.
Cornish hens,				
1 to 1½ pounds each	(1) 12 to 15	DEFROST	20 minutes	
	(2) 20 to 25	DEFROST	20 minutes	

Poultry—How to Cook

1. Defrost poultry completely before cooking.
2. Remove giblets; rinse poultry in cool water.
3. When cooking whole birds, place on a microwave roasting rack or inverted saucer in glass baking dish large enough to catch drippings.
4. Cook whole poultry uncovered. Toward the end of the cooking, small pieces of foil may be used to cover legs, wing tips, or breast bone area to prevent overcooking. Foil should be at least 1 inch away from oven walls.
5. When cooking poultry pieces, arrange in glass baking dish with thick edges toward outside of dish.
6. Cover poultry pieces with either glass lid or plastic wrap during cooking and standing time.
7. Use microwave meat thermometer to test for doneness at the end of cooking. Doneness in all poultry is determined when the meat cut near the bone is no longer pink and the thermometer registers 170°F. In whole birds, temperature

reading is most accurate when thermometer is inserted into the thickest part of the thigh. A regular meat thermometer should *not* be used in the microwave oven. However, it can be used to test doneness (170°F.) when poultry is removed from oven.

8. Standing time completes the cooking of poultry. Cover cooked whole birds with foil during standing time.

Poultry	Minutes per Pound	Setting	Standing Time	Special Notes
Turkey (whole), 8 to 14 pounds	9	ROAST	10 to 15 minutes	Start breast side down; turn breast side up at halfway point in cooking.
Turkey breast	7 5	HIGH ROAST	10 to 15 minutes	Begin on HIGH; turn over; continue on ROAST.
Turkey drumsticks	15	ROAST	5 minutes	
Chicken (whole): 2 to 3 pounds 3 to 5 pounds	7 8	HIGH HIGH	5 minutes 5 minutes	Protect ends of legs and wings with pieces of foil.
Chicken (quartered), 2 to 3 pounds	7	HIGH	5 minutes	Turn pieces over when half cooked.
Chicken (cut-up)		HIGH	5 minutes	Turn pieces over when half cooked.
Duckling, 4 to 5 pounds	8	ROAST	10 to 15 minutes	Turn over halfway through cooking and drain off excess fat. For crisper skin, place duckling under conventional oven broiler before serving.
Cornish hens, 1 to 1½ pounds each	(1) 8 (2) 9	HIGH HIGH	5 minutes 5 minutes	Turn over halfway through cooking time.

Quick Baked Chicken   4 servings

2½- to 3-pound broiler-fryer
  chicken
Salt and pepper
1 small onion, quartered

2 stalks celery, cut in chunks
2 tablespoons soft butter or
  margarine
⅛ teaspoon thyme

1. Wash chicken. Remove giblets. Sprinkle inside of body cavity with salt and pepper. Place onion and celery inside body cavity. Tie legs together with string. Tie wings to body of chicken.
2. Place chicken, breast side up, on a roasting rack or inverted saucer in a 12- by 7-inch glass baking dish. Spread with soft butter and sprinkle with thyme.
3. Cook chicken on HIGH for 15 to 17 minutes.
4. Doneness is tested by inserting temperature probe in thickest part of thigh at end of cooking. Chicken is done when temperature reaches 170°F. A microwave thermometer or a conventional thermometer can be used after the poultry is removed from the oven to test for doneness.
5. Let stand, covered with foil, about 5 minutes before serving.

Chicken Cacciatore

4 servings

¼ cup cooking oil
1 broiler-fryer chicken, cut in serving pieces
1 medium onion, coarsely chopped
1 clove garlic, minced
1 medium green pepper, seeded and coarsely chopped

1¼ teaspoons salt
⅛ teaspoon pepper
½ bay leaf
1 can (16 ounces) tomatoes
2 tablespoons dry white wine
Chopped parsley

1. Pour oil into an 8-inch square baking dish. Cut large pieces of chicken in half. Put in baking dish and turn so that all pieces are coated with oil. Place breasts in center and surround with other pieces of chicken. Sprinkle onion and garlic on top of chicken. Cook on HIGH for 5 minutes.
2. Combine green pepper, salt, pepper, bay leaf, tomatoes, and wine. Stir with a fork and mash tomatoes into small pieces. Pour over top of chicken.
3. Cook, covered, on HIGH for 23 to 25 minutes, or until chicken is tender.
4. Garnish with parsley and serve with cooked spaghetti.

Bacon-Flake Chicken

4 to 6 servings

5 slices bacon, cut in half
3 whole chicken breasts
1 cup milk

2 cups crushed corn flakes
Salt and pepper
Garlic salt

1. Put bacon in a shallow 8- by 10-inch baking dish. Cover with waxed paper and cook on HIGH for 5 minutes.
2. Turn bacon over. Cook on HIGH for 3 minutes, or until very crisp.
3. Cut chicken breasts in half. Remove skin and breastbone. Dip chicken in milk and roll in corn flakes. Pat corn flakes on so that they stick all over chicken. Place on top of cooked bacon. Pour remaining milk carefully into dish without disturbing chicken.
4. Cook, uncovered, on HIGH for 5 minutes.
5. Turn chicken over. Sprinkle with salt and pepper. Cook on BAKE for 8 minutes. Let stand 3 minutes before serving. Sprinkle with garlic salt just before serving.

Barbecued Chicken

4 servings

2½- to 3-pound broiler-fryer chicken, quartered
½ cup bottled barbecue sauce

1 tablespoon dried parsley flakes
1 tablespoon dried onion flakes

1. place chicken pieces, skin side up, with thick edges toward outside of a 12- x 7-inch glass baking dish.
2. Combine remaining ingredients. Brush half of mixture over top of chicken.
3. Cook, covered, on HIGH for 10 minutes. Brush with remaining barbecue sauce.
4. Cook, covered, on HIGH for 8 to 10 minutes, or until chicken is tender.
5. Let stand, covered, 5 minutes before serving.

Cheesy Chicken

4 servings

¼ cup butter or margarine
¾ cup saltine cracker crumbs
⅓ cup grated Parmesan cheese
2 tablespoons chopped chives

2 tablespoons chopped parsley
1 broiler-fryer chicken, about 3 pounds, cut in serving pieces

1. Place butter in an 8-inch square glass baking dish. Cook on ROAST for 1½ minutes, or until butter is melted.
2. Combine cracker crumbs, cheese, chives, and parsley in a shallow dish.
3. Dip pieces of chicken in butter and roll in cracker mixture.
4. Place chicken pieces in 8-inch square glass baking dish, skin side up, with thickest pieces of chicken toward outside of dish.
5. Cook, covered, on HIGH for 15 to 17 minutes, or until chicken is tender.
6. Let stand, covered, about 5 minutes before serving.

Coq Au Vin

4 servings

1 large onion, sliced
1 clove garlic, minced
1 teaspoon instant chicken bouillon
1 teaspoon salt
2 tablespoons minced parsley

¼ teaspoon thyme
¾ cup dry red wine
2½ to 3-pound broiler-fryer chicken, cut in serving pieces
½ pound whole mushrooms

1. Combine onion, garlic, bouillon, salt, parsley, thyme, and wine in a 2-quart casserole or baking dish. Add chicken pieces.
2. Cook, covered, on HIGH for 20 minutes.
3. Turn chicken pieces over. Add mushrooms. Cook, covered, on ROAST for 10 minutes.
4. Let stand 5 minutes before serving.

Lemon Chicken Wings

28 pieces

3 pounds chicken wings
½ cup salad oil
½ cup lemon juice

1 clove garlic, crushed
1 teaspoon salt
½ teaspoon pepper

1. Cut chicken wings apart at both joints. Reserve wing tips for soup stock.
2. Combine remaining ingredients in a 12- by 7-inch glass baking dish. Add chicken pieces and let stand about 1 hour, turning chicken pieces several times.
3. Cook, covered, on HIGH for 6 minutes. Turn chicken wings over and cook on HIGH for 6 to 7 minutes, or until chicken is tender.
4. Let stand, covered, 5 minutes before serving.

Note: Serve as finger food at a cocktail party or as a main course for dinner.

Nut Stuffing for Poultry

1 cup butter or margarine, divided
½ cup salad oil
2 large onions, minced
4 stalks celery, minced
5 quarts day-old bread crumbs or ½-inch squares

2 teaspoons poultry seasoning
¼ cup chopped parsley
2 teaspoons salt
1 cup coarsely chopped walnuts or pecans

1. Place ½ cup butter and salad oil in a 1½-quart casserole or baking dish. Add onions and celery. Cook, covered, on SIMMER for 5 to 6 minutes. Combine with bread crumbs, poultry seasoning, parsley, salt, and nuts. Toss lightly.
2. Wash completely thawed turkey. Stuff neck opening with part of the stuffing. Secure skin flap with strong toothpicks or wooden skewers. Stuff body cavity with remaining stuffing. Tie drumsticks together with strong string. Tie wings tightly to body with string.
3. Cook turkey according to poultry cooking chart.

Note: Makes enough stuffing for a 12-pound turkey.

Turkey Tetrazzini   6 servings

4 ounces thin spaghetti
3 tablespoons butter or margarine
1 can (4 ounces) sliced mushrooms, drained
⅓ cup finely minced onions
3 tablespoons all-purpose flour
2 cups chicken broth or milk

½ cup light cream
¼ cup dry vermouth
1 teaspoon salt
Dash of white pepper
¾ cup grated Parmesan cheese, divided
2 cups diced cooked turkey

1. Cook spaghetti according to package directions. Drain immediately and rinse in cold water to stop cooking. Reserve.
2. Place butter in a 3-quart casserole. Add mushrooms and onion.
3. Cook, covered, on SAUTE for 3 to 4 minutes, or until onions are soft.
4. Add flour and mix to form a smooth paste.
5. Cook, covered, on HIGH for 30 seconds.
6. Stir in chicken broth, cream, vermouth, salt, pepper, and ¼ cup Parmesan cheese. Blend well.
7. Cook, uncovered, on HIGH for 4 to 5 minutes, or until mixture comes to a boil and thickens. Stir once during cooking period.
8. Add cooked spaghetti, turkey, and remainder of cheese. Toss slightly.
9. Cook, covered, on BAKE for 8 to 9 minutes, or until piping hot.
10. Let stand on WARM for 5 minutes before serving.

Note: Chicken or ham may be substituted for the turkey.

Turkey Divan

4 to 6 servings

1 bunch broccoli, cooked
8 to 12 slices cooked turkey breast
2 cups white sauce (see page 108)

¼ cup grated Parmesan cheese
¼ cup grated Gruyere cheese

1. Place cooked broccoli in a baking dish with flower ends at ends of dish. Cover center of stalks with slices of cooked turkey.
2. Prepare white sauce, and while it is still warm add the cheeses. Stir until well blended. Pour over top of turkey. Sprinkle with additional Parmesan cheese if desired.
3. Cook, covered, on BAKE for 9 to 10 minutes, or until turkey is piping hot.

Oriental Turkey Drumsticks

2 to 3 servings

2 turkey drumsticks, about 2 to 3 pounds
¼ cup dry sherry

2 tablespoons soy sauce
1 clove garlic, minced

1. Place turkey drumsticks, with thick ends toward the outside, in a 12- by 7-inch glass baking dish.
2. Combine remaining ingredients. Pour over drumsticks. Let stand, turning occasionally, about 30 minutes.
3. Cook, covered, on ROAST for 20 minutes. Turn drumsticks over and baste with juice in pan. Cook, covered, on ROAST for 15 to 20 minutes, until tender, or until meat cut near the bone is no longer pink.
4. Let stand, covered, 5 minutes before serving.

Cornish Hen for Two

2 servings

2 Cornish hens, 1 to 1½ pounds each
2 cups stuffing (optional)

½ cup currant jelly
1 tablespoon dry sherry

1. Wash hens. Remove giblets and set aside. Fill hens with stuffing, if desired. Tie legs together. Tie wings to body.
2. Place hens, breast side down, on a roasting rack or inverted saucer in a 12- by 7-inch glass baking dish.
3. Put jelly and sherry in a 1-cup glass measuring cup. Cook on HIGH for 1 minute, or until jelly softens. Stir well. Brush ½ of mixture over top of hens.
4. Cook on HIGH for 15 minutes. Turn hens breast side up. Brush with remaining jelly mixture. Cook on HIGH for 10 minutes, or until meat cut near bone is no longer pink.
5. Let stand, covered with foil, 5 minutes before serving.

Turkey Divan

Roast Orange Duckling

4 servings

4- to 5- pound duckling, thawed
1 orange, peeled and cut in chunks

1 medium onion, quartered
½ cup orange marmalade

1. Wash duckling and remove giblets. Place orange and onion pieces in body cavity of duckling. Secure neck skin flap with strong toothpicks. Tie legs together and tie wings to body.
2. Place duckling, breast side down, on a roasting rack or inverted saucer in a 12-by 7-inch glass baking dish.
3. Put marmalade in a small glass bowl. Cook on SIMMER for 4 minutes.
4. Spread warm marmalade on duckling. Cook on ROAST for 20 minutes.
5. Remove from oven and drain off excess fat. Turn breast side up. Brush with remaining marmalade. Cook, covered, on ROAST for 15 minutes, or until meat near bone is no longer pink. Remove duckling from oven and test for temperature of 170°F.
6. Let stand, covered with foil, 10 minutes.

Note: For a crisp skin, brown duckling under a hot broiler until brown and crispy on all sides. Then cover with foil and let stand a few minutes.

Fish

The microwave oven is one of the most convenient time and flavor-saving methods of cooking fish. Since fish and shellfish are delicate, the cooking should be done quickly in order to retain flavor and moisture; it is better to undercook fish so that it will not dry out. Arrange large pieces and thick edges of fish toward the outside edges of the baking dish. Most fish is cooked with a cover. Plastic wrap, tightly tucked over the baking dish, will do the job; the dish should remain covered during the standing time. Pay special attention to the cooking of shellfish—the ROAST setting is recommended—as it will toughen rapidly if overcooked.

When cooking meals of fish with vegetables or other foods, *always* cook the fish last. Cook until the fish flakes easily when tested with a fork.

Fish—How to Defrost

1. Fish may be thawed in the original packaging.
2. Place fish in a flat, glass baking dish for defrosting.
3. Use DEFROST setting to thaw fish.

Fish	Time in Minutes	Setting	Standing Time in Minutes	Special Notes
Fish fillets (1 pound)	10 to 12	DEFROST	5	Carefully separate fillets under cold running water.
Lobster tails (1 pound)	10 to 12	DEFROST	5	If two in a package, separate under cold running water. Turn over half-way through defrosting time.
Shrimp (1 pound)	5 to 6	DEFROST	5	Separate shrimp halfway through defrosting time.
Whole fish (1½ to 2 pounds)	13 to 14	DEFROST	5	Carefully rinse in cold water to finish defrosting.
Fish steaks (1 pound)	6 to 7	DEFROST	5	Carefully separate steaks under cold running water.

Fish—How to Cook

1. Fish should be thoroughly thawed before cooking.
2. Remove original wrappings and rinse under cold running water.
3. If fish is not thoroughly defrosted, thawing can be completed easily under cold running water.
4. Cook fish in a glass baking dish or platter. Place fish fillets and steaks in container with thickest edges of fish toward outside of dish. Cover with glass lid or plastic wrap to retain moisture.
5. Cook fish quickly on HIGH just until it flakes easily when tested with a fork.

Fish	Time in Minutes	Setting	Standing Time in Minutes
Fish fillets (1 pound)	5 to 6	HIGH	4 to 5
(2 pounds)	7 to 8	HIGH	4 to 5
Lobster tails (1 pound)	7 to 8	HIGH	3 to 4
Shrimp (1 pound)	5 to 7	ROAST	4
Whole fish (1½ to 2 pounds)	10 to 13	ROAST	4 to 5
Fish steaks (1 pound)	5 to 7	HIGH	5 to 6

Shrimp Creole

6 servings

3 tablespoons butter or margarine	1 tablespoon Worcestershire sauce
½ cup chopped onions	1½ teaspoons salt
½ cup chopped green pepper	1 teaspoon sugar
½ cup diced celery	½ teaspoon chili powder
1 clove garlic, minced	Dash of hot-pepper sauce
1 can (1 pound) tomatoes, mashed	1 tablespoon cornstarch
1 can (8 ounces) tomato sauce	1 pound cooked shrimp

1. Combine butter, onion, green pepper, celery, and garlic in a 2- or 3-quart casserole.
2. Cook, uncovered, on HIGH for 3 minutes.
3. Add tomatoes, tomato sauce, Worcestershire, salt, sugar, chili powder, and hot-pepper sauce.
4. Cook, uncovered, on HIGH for 7 minutes, stirring once.
5. Combine cornstarch with 2 teaspoons cold water. Stir into casserole.
6. Cook, uncovered, on HIGH for 3 minutes, stirring once.
7. Add shrimp. Cook, uncovered, on HIGH for 2 minutes, or until shrimp is piping hot.
8. Serve with wild rice or plain boiled rice.

Shrimp Jambalaya

4 to 6 servings

¼ cup butter or margarine	¾ cup chicken bouillon
1 onion, chopped	1 teaspoon salt
½ cup chopped green pepper	½ teaspoon hot pepper sauce
1 clove garlic, minced	1 pound cooked, peeled shrimp
1 cup raw rice	1½ cups diced cooked boiled or
1 can (16 ounces) tomatoes	baked ham

1. Place butter, onion, green pepper, and garlic in a 2½- to 3-quart glass baking dish or casserole. Cook, covered, on HIGH for 5 minutes, or just until vegetables are soft.
2. Add rice, tomatoes, bouillon, salt, and hot pepper sauce. Stir. Cook, covered, on HIGH just until mixture comes to a boil. Cook on SIMMER for 20 minutes.
3. Stir in shrimp and ham. Cook, covered, on SIMMER for 10 minutes.
4. Let stand 5 minutes before serving.

Note: In place of ham, use 1½ cups diced, cooked turkey or chicken.

Fish Fillets with Mushrooms

3 to 4 servings

1 pound fish fillets	1 tomato, peeled and cut into
2 tablespoons butter or	pieces
margarine	½ teaspoon salt
2 green onions, thinly sliced	½ tablespoon lemon juice
½ cup fresh sliced mushrooms	2 tablespoons dry white wine

1. Arrange fish fillets in a 13- by 9-inch glass baking dish with the thick edges toward the outside of the dish. Dot with butter. Sprinkle remaining ingredients over top of fish.
2. Cook, covered, on HIGH for 5 minutes.
3. Let stand, covered, 5 minutes.

Fillet of Sole Almondine

4 servings

½ cup slivered almonds
½ cup butter or margarine
1 pound fresh or frozen fillet of sole
½ teaspoon salt

⅛ teaspoon pepper
1 teaspoon chopped parsley
1 tablespoon lemon juice

1. Place almonds and butter in an 8-inch square baking dish.
2. Cook, uncovered, on HIGH for 5 minutes, or until almonds and butter are golden brown. Remove almonds with a slotted spoon and set aside.
3. Arrange sole in dish with butter, turning to coat both sides of fish. Sprinkle with salt, pepper, parsley, and lemon juice.
4. Cook, covered with waxed paper, on HIGH for 4 minutes.
5. Remove waxed paper. Sprinkle fish with toasted almonds. Cook, covered, on HIGH for 2 minutes, or until fish flakes easily when tested with a fork.
6. Let stand 1 to 2 minutes before serving. To serve, garnish with lemon wedges and sprigs of parsley.

Stuffed Bass

4 to 6 servings

1 whole bass, cleaned, about 3 pounds
2 cups fresh bread cubes
1 tablespoon melted butter or margarine
Salt and pepper to taste

1 tablespoon chopped parsley
1 teaspoon lemon juice
3 slices bacon

1. Wash fish in cold water. Pat dry with paper towels.
2. Combine bread cubes, butter, salt and pepper, parsley, and lemon juice. Toss lightly. Stuff cavity of fish. Close with toothpicks or tie with string.
3. Place fish in an oblong baking dish. Place bacon slices on fish.
4. Cook, covered, on HIGH for 16 to 18 minutes, or until fish flakes easily when tested with a fork.
5. Let fish stand covered with paper towels 5 minutes before serving.

Halibut Steaks

4 servings

2 halibut steaks, about ¾ inch thick
½ lemon
1 egg, beaten
½ can (10¾ ounces) condensed cream of celery soup

2 tablespoons milk
2 tablespoons grated Parmesan cheese, divided
2 tablespoons fine dry bread crumbs
2 teaspoons melted butter or margarine

1. Wipe fish with a damp paper towel. Cut each steak in 2 portions. Place in an 8-inch square baking dish. Squeeze lemon over top of fish. Set aside.
2. In a 2-cup measure beat together egg, soup, milk, and 1 tablespoon cheese.
3. Cook, covered, on ROAST for 2 minutes. Remove and stir well to melt cheese.
4. Pour soup mixture over steaks. Combine bread crumbs and melted butter and sprinkle over top of fish. Top with remaining 1 tablespoon cheese.
5. Cook, covered, on HIGH for 6 minutes, or until fish flakes when tested with a fork.

Fillet of Sole Almondine

Tuna Crunch

4 servings

1 cup thinly sliced celery
¼ cup chopped onion
2 tablespoons butter or margarine
1 can (7 ounces) tuna fish

1 can (10¾ ounces) cream of
mushroom soup, undiluted
1 can (3 ounces) chow mein noodles
½ cup coarsely chopped cashews

1. Combine celery, onion, and butter in a 1-quart casserole. Cook, uncovered, on HIGH for 5 minutes, stirring once during cooking time.
2. Drain tuna and flake with a fork. Add to onion mixture with soup, ⅔ can of noodles, and cashews. Toss lightly.
3. Cook, covered, on HIGH for 2 minutes.
4. Stir mixture lightly. Top with remaining noodles. Cook, covered, on HIGH for 3 to 4 minutes, or until piping hot.

Tuna-Spinach Casserole

4 servings

1 package (10 ounces) raw spinach
1 can (7 ounces) solid pack tuna
1 can (4 ounces) sliced mushrooms
2 tablespoons lemon juice
3 tablespoons butter or margarine, divided

1 tablespoon minced onion
2 tablespoons all-purpose flour
½ teaspoon salt
⅛ teaspoon pepper
1 egg, lightly beaten

1. Rinse spinach in fresh, cold water. Drain well. Break in pieces, removing tough center stems. Put in a 2-quart casserole.
2. Cook, covered, on HIGH for 3 to 4 minutes, or until spinach is limp. Drain well and set aside.
3. Drain tuna and set aside.
4. Drain mushrooms, reserving liquid. Put mushroom liquid in a 1-cup measure. Add lemon juice and enough water to make 1 cup of liquid.
5. Put 2 tablespoons of the butter in a 1-quart casserole. Cook on HIGH for 30 seconds, or just long enough to melt butter.
6. Add onion, flour, salt, and pepper. Cook, uncovered, on BAKE for 1 minute.
7. Stir in mushroom liquid. Cook, uncovered, on BAKE for 5 minutes, or until thick, stirring occasionally during cooking time.
8. Add a small amount of sauce to egg, beat well, and return to hot sauce. Stir mushrooms into sauce.
9. Put drained spinach in a 2- to 3-quart casserole. Break tuna in big chunks and place over top of spinach. Pour sauce over top. Dot with remaining 1 tablespoon of butter.
10. Cook, uncovered, on HIGH for 4 minutes. Let stand covered with waxed paper 3 to 5 minutes before serving.

Scallops Cacciatore

4 servings

1 medium onion, chopped
1 medium green pepper, chopped
¼ cup salad oil
1 can (1 pound) tomatoes, drained
1 pound bay scallops
1 can (8 ounces) tomato sauce

¼ cup dry white wine
1¼ teaspoons salt
⅛ teaspoon pepper
2 bay leaves
2 tablespoons chopped parsley

1. Combine onion, green pepper, and oil in a 1½- to 2-quart casserole.
2. Cook, covered, on HIGH for 3 minutes, stirring once.
3. Mash tomatoes with a fork to break into small pieces. Add to casserole with scallops, tomato sauce, wine, salt, pepper, and bay leaves.
4. Cook, covered, on HIGH for 6 minutes, or until scallops are tender.
5. Sprinkle with parsley. Serve with hot cooked rice, if desired.

Scallops Poulette

4 servings

¼ cup butter or margarine
1 tablespoon minced onion
2 tablespoons flour
1 can (4 ounces) sliced mushrooms,
   drained
¼ cup dry vermouth
½ teaspoon salt

⅛ teaspoon pepper
1 pound bay scallops
1 bay leaf
2 teaspoons lemon juice
½ cup light cream
1 egg yolk
1 tablespoon chopped parsley

1. Combine butter and onion in a 2-quart casserole.
2. Cook, uncovered, on HIGH for 2 minutes.
3. Stir in flour and blend well. Add mushrooms, wine, salt, pepper, scallops, bay leaf, and lemon juice. Toss together lightly.
4. Cook, covered, on HIGH for 6 minutes, or until scallops are tender.
5. Remove bay leaf. Beat together cream and egg yolk. Add some hot liquid carefully to egg and blend well. Stir egg mixture carefully into hot casserole. Stir well.
6. Cook, covered, on BAKE for 5 minutes, stirring once during cooking period.
7. Sprinkle with parsley and serve.

Steamed Clams

2 servings

1 quart steamer clams   **Melted butter or margarine**

1. Scrub clams with a stiff brush to remove all sand and grit. Discard all clams that are even the least bit open.
2. Put scrubbed clams in a 2-quart casserole. Add 2 tablespoons water.
3. Cook, covered, on HIGH for 8 minutes, or until clam shells are open and clams are cooked. Discard clams with closed shells.
4. Serve clams in flat bowls. Divide clam liquid into custard cups and fill a second set of cups with melted butter. Remove clams from shells, dip in clam liquid, and then in butter.
5. Drink clam broth when clams have been eaten.

Spicy Clams

2 servings

2 tablespoons butter or margarine
1 medium onion, chopped
2 cloves garlic, finely minced
1½ cups clam-tomato juice cocktail

½ teaspoon oregano
Dash liquid hot pepper sauce
24 clams, well scrubbed and washed
Chopped parsley

1. Place butter, onion, and garlic in an 11- by 7-inch glass baking dish. Cook on HIGH for 3 minutes. Add clam-tomato juice, oregano, and hot pepper sauce. Cook on HIGH for 3 to 4 minutes, or until mixture comes to a boil.
2. Arrange clams in a single layer. Cook, covered, on HIGH for 8 minutes, or until shells pop open.
3. Sprinkle with chopped parsley, and serve in soup plates with the broth.

Lobster Tails

4 servings

1 pound frozen lobster tails
1 teaspoon lemon juice

¼ cup butter or margarine
¼ teaspoon grated lemon peel

1. Place frozen lobster tails in an 8-inch square baking dish. Sprinkle with lemon juice. Cook on HIGH for 4 minutes to thaw lobster.
2. Cut away soft shell-like surface on underside of tail with a sharp knife or scissors. Insert wooden skewers into each lobster tail to keep flat while cooking. Drain liquid from dish and arrange lobster in dish shell side down. down.
3. Combine butter and lemon peel in a small custard cup. Cook on ROAST for 30 seconds, or until butter is melted.
4. Brush tails with butter mixture.
5. Cook, covered, on HIGH for 7 minutes, or until lobster meat turns pink.
6. Serve hot with melted butter and lemon wedges.

Steamed Clams

Crab Meat in Shells
4 servings

1 tablespoon chopped parsley
1 tablespoon chopped green pepper
1 scallion, chopped, including green top
1 can (4 ounces) sliced mushrooms and pieces, well drained
1 teaspoon butter
1 can (10¾ ounces) condensed cream of celery soup, undiluted

1 can (7½ ounces) crab meat
1 tablespoon lemon juice
1 tablespoon dry sherry
2 tablespoons dry bread crumbs
2 tablespoons grated Cheddar cheese
Paprika

1. Combine parsley, green pepper, scallion, mushrooms, and butter in a 1-quart casserole.
2. Cook, covered, on HIGH for 3 minutes.
3. Add soup and blend well. Cook, covered, on HIGH for 1 minute.
4. Pick over crab meat and remove any bits of cartilage. Add to mixture with lemon juice and sherry. Divide mixture into 4 scallop shells. Combine bread crumbs and cheese. Spread over top of crab mixture. Sprinkle generously with paprika.
5. Cook two shells at a time on HIGH for 3 minutes, or until piping hot.

Fish Sticks with Bearnaise Sauce
4 servings

¼ cup butter or margarine
1 tablespoon finely minced onion
¼ cup light cream
1 tablespoon lemon juice
1 tablespoon dry white wine

2 egg yolks, lightly beaten
½ teaspoon dry mustard
¼ teaspoon salt
½ teaspoon crushed tarragon leaves
1 package (16 ounces) fish sticks

1. Place butter and onion in a 1-quart glass measure. Cook on ROAST for 1½ minutes, or until butter is melted.
2. Using a rotary beater, beat in cream, lemon juice, white wine, and egg yolks. Add mustard, salt, and tarragon leaves, and continue beating until blended.
3. Cook on ROAST for 1 minute. Remove from oven and beat thoroughly with a rotary beater. Cook on ROAST for 1 to 1½ minutes, or until mixture is thickened. Remove from oven and beat lightly.
4. Place fish sticks in a single layer in a 3-quart glass baking dish. Cook on HIGH for 10 to 12 minutes, or until thoroughly heated.
5. Serve immediately with the sauce.

Eggs
and Cheese

Eggs are always cooked on the ROAST setting. It is wise to prick the yolk gently with the point of a sharp knife—just enough to allow the steam to escape while they are cooking so that the yolks will not burst. The yolks tend to cook faster than the whites. Eggs should *never* be cooked in the shell.

Cheese, too, cooks best on the ROAST setting. Within their own category—soft, semi-soft, hard, or processed—cheeses can be interchanged in any recipe. Undercook slightly, rather than overcook any cheese dish.

Since eggs and cheese both fare best on ROAST, many superb dishes that combine both can be prepared in a microwave oven.

Eggs — How to Scramble

1. Break eggs into a bowl or use a 1-quart glass casserole for 4, 5, or 6 eggs.
2. Add milk or cream and beat with a fork.
3. Add butter and season to taste.
4. Cover with a lid, a small plate, or waxed paper.
5. Cook on BAKE. Stir once during cooking period.
6. Let stand 1 minute before serving.

Number of Eggs	Liquid	Butter	Minutes to Cook
1	1 tablespoon	1 teaspoon	1 to 1½
2	3 tablespoons	2 teaspoons	2 to 2½
4	¼ cup	4 teaspoons	4½ to 5½
6	6 tablespoons	2 tablespoons	7 to 8

Eggs — How to Poach

1. Use a 6-ounce custard cup or coffee cup when poaching 1, 2, or 3 eggs. Poach 4 eggs in a 1-quart casserole.
2. Bring water to a boil with a pinch of salt and ¼ teaspoon vinegar on HIGH.
3. Break eggs carefully into hot water.
4. Cover with a lid, a small plate, or waxed paper.
5. Cook on SIMMER.
6. Let stand, covered, 1 minute before serving to allow egg whites to set.

Number of Eggs	Water	Container	Minutes to Boil Water	Minutes to Cook
1	¼ cup	6-ounce custard cup	1½ to 2	1
2	¼ cup	6-ounce custard cups	2	1½ to 2
4	1 cup	1-quart casserole	2½ to 3	2½ to 3

Shirred Eggs   1 serving

1 teaspoon butter or margarine   **1 tablespoon cream**
2 eggs

1. Melt butter in a ramekin or small cereal bowl on ROAST for 30 seconds.
2. Break eggs carefully into ramekin or bowl. Add cream. Cover tightly with plastic wrap.
3. Cook, covered, on BAKE for 2 to 2½ minutes.
4. Remove and let stand 1 minute before serving.

Shirred Eggs and Sticky Buns (recipe, p. 138)

Eggs Benedict

4 servings

¾ cup hollandaise sauce

2 English muffins, split and toasted

4 slices ham, ¼ to ½ inch thick

4 poached eggs

1. Prepare hollandaise sauce. Cover with a piece of waxed paper and set aside.
2. Place each muffin half on a paper plate and top each with 1 slice ham.
3. Cook, uncovered, 2 at a time, on ROAST for 2 to 2½ minutes, or until ham is hot.
4. Top each with a poached egg. Cover with hollandaise sauce and serve immediately.

Omelet

1 to 2 servings

1 tablespoon butter or margarine

3 eggs

3 tablespoons water

½ teaspoon salt

⅛ teaspoon pepper

1. Place butter in a 9-inch glass pie plate. Cook on HIGH for 45 seconds, or until butter is melted.
2. Beat remaining ingredients lightly with a fork. Pour into pie plate. Cook, covered, on ROAST for 3 minutes. Stir lightly. Cook, covered, on ROAST for 1½ to 2 minutes, or until almost set in center.
3. Let stand, covered, 1 to 2 minutes before serving. Fold in half and serve.

Variations: Before folding omelet, top with crumbled cooked bacon, grated Cheddar cheese, chopped ham, or chopped tomato.

Western Omelet

Fold into eggs before cooking:

¼ cup finely chopped cooked ham

1 tablespoon finely chopped green onion

1 tablespoon minced onion

Poached Eggs in Corned Beef Hash

4 servings

1 tablespoon butter or margarine

1 large onion, finely chopped

2 cups chopped or ground cooked corned beef

3 cups finely chopped cooked potatoes

1 can (8 ounces) cooked beets, drained and finely chopped

½ cup milk

⅓ cup catsup

4 eggs

1. Place butter and onion in an 8-inch square glass baking dish. Cook on HIGH for 3 minutes, or until onion is partially cooked.
2. Stir in corned beef, potatoes, beets, milk, and catsup. Cover with a piece of waxed paper. Cook on ROAST for 10 minutes, or until mixture is hot.
3. Make 4 indentations in hash with the back of a spoon. Break an egg into each indentation. Cook, covered, on ROAST for 5 minutes, or just until eggs are set.
4. Let stand 2 to 3 minutes before serving.

74

Cheesed Ham and Eggs

4 to 6 servings

¼ cup butter or margarine
¼ cup all-purpose flour
2 cups milk
1½ teaspoons prepared mustard
2 teaspoons Worcestershire sauce

1 cup shredded Cheddar cheese
1 cup diced cooked ham
6 hard-cooked eggs, peeled and halved
4 to 6 slices of bread, toasted

1. Place butter in a 1½-quart casserole or baking dish. Cook on HIGH for 1 minute, or until butter melts. Blend in flour. Stir in milk. Cook on HIGH for 4 to 6 minutes, stirring once during cooking time.
2. Stir with a wire wisk. Add mustard, Worcestershire sauce, and cheese. Cook on HIGH for 30 seconds, or until cheese is melted.
3. Stir in ham. Carefully fold in eggs. Cook on HIGH for 1½ to 2 minutes, or until piping hot.
4. Serve over toast.

Hollandaise Sauce

¾ cup

¼ cup butter or margarine
¼ cup light cream
2 egg yolks, well beaten

1 tablespoon lemon juice
½ teaspoon dry mustard
¼ teaspoon salt

1. Place butter in a 1-quart glass measure. Cook on HIGH for 1 minute, or until butter is melted.
2. Add remaining ingredients. Beat with a rotary beater until smooth.
3. Cook on ROAST for 1 minute. Beat well and cook on ROAST for 1 minute, or until thickened. Remove and beat with rotary beater until light and smooth.

Note: If sauce curdles, beat in 1 teaspoon water and continue beating until mixture is smooth.

Cheese Strata

6 servings

8 slices day-old bread
1 cup shredded Cheddar cheese
4 eggs
2½ cups milk

1 small onion, finely minced
½ teaspoon prepared mustard
1 teaspoon salt
⅛ teaspoon pepper
Paprika

1. Trim crusts from bread. Place 4 slices of bread in the bottom of an 8-inch square glass baking dish. Cover with shredded cheese. Top with remaining 4 slices of bread.
2. Beat eggs lightly. Beat in milk, onion, mustard, salt, and pepper. Pour over top of bread slices. Sprinkle with paprika. Cover with waxed paper and let stand at room temperature at least 1 hour.
3. Cook on BAKE for 25 to 28 minutes, or until a knife inserted in center comes out clean.
4. Let stand 3 minutes before serving.

Welsh Rabbit on Toast

4 to 6 servings

4 teaspoons butter or margarine

4 cups (1 pound) shredded sharp
Cheddar cheese

¾ teaspoon Worcestershire sauce

½ teaspoon salt

½ teaspoon paprika

¼ teaspoon dry mustard

¼ teaspoon cayenne

2 eggs, lightly beaten

1 cup flat beer or ale, at room
temperature

1. Melt butter in a 2-quart casserole or bowl on BAKE for 2 minutes.
2. Add cheese, Worcestershire, salt, paprika, dry mustard, and cayenne. Mix thoroughly.
3. Cook, covered, on SIMMER for 6 minutes, stirring once during cooking time.
4. Stir a little of the hot cheese into beaten eggs. Return slowly to hot mixture and stir briskly. Gradually stir in beer and blend well.
5. Cook, covered, on SIMMER for 3 minutes. Remove from oven and stir well.
6. Cook, covered, on SIMMER for 3 minutes. Remove from oven and beat briskly with a whisk to blend thoroughly.
7. Serve over crisp toasted French bread slices and garnish with tomatoes and crisp bacon slices.

Sunday-Night Special

4 to 6 servings

1 can (7 ounces) green chilies

1 cup coarsely crushed corn chips

Meat (optional)

2 mild Italian sausages, cooked
and broken up, or

½ pound ground chuck, cooked, or

1 cup ground cooked pork or ham

½ cup cottage cheese or ricotta

4 ounces Monterey Jack cheese, cut in
strips

2 eggs

1 cup milk

½ teaspoon salt

½ cup grated Cheddar or Parmesan
cheese

1. Wash and dry chilies. Remove any remaining seeds. Cut into strips 1 inch wide.
2. Put half the corn chips in bottom of an 8-inch round cake pan. Arrange about one-third of the chilies on top of corn chips. Place meat, if desired, on top of chilies. Dot cottage cheese on top of meat. Add another third of the chilies. Arrange Monterey Jack cheese over top of chilies. Arrange remaining chilies on top of cheese.
3. Beat together eggs, milk, and salt. Pour over top of mixture in casserole. Sprinkle top with Cheddar or Parmesan cheese. Sprinkle remaining corn chips on top of cheese.
4. Cook, uncovered, on BAKE for 16 to 18 minutes, or until custard is set.
5. Cook, covered, on LOW for 5 minutes.

Swiss Cheese Fondue

6 servings

4 cups shredded Swiss cheese	**Dash of pepper**
¼ cup all-purpose flour	**2 cups dry white wine**
¼ teaspoon salt	**2 tablespoons Kirsch**
¼ teaspoon nutmeg	**1 loaf French bread, cut into cubes**

1. In a 1½-quart dish or casserole combine cheese, flour, salt, nutmeg, and pepper. Toss lightly to coat cheese with flour. Stir in wine.
2. Cook, covered, on SIMMER for 6 minutes, stirring three times during cooking time. Stir well after removing from oven to finish melting cheese.
3. If cheese is not all melted, cook, covered, on SIMMER for 1 minute.
4. Stir in Kirsch.
5. Serve immediately with cubes of French bread. Spear each cube of bread on a fondue fork; dip into fondue and eat immediately.
6. If fondue cools during eating time, rewarm on SIMMER for 1 to 2 minutes.

Swiss and Onion Pie

6 to 8 slices

4 slices bacon	**1 tablespoon all-purpose flour**
1 large onion, thinly sliced	**3 eggs, lightly beaten**
1 tablespoon butter or margarine	**1 cup milk**
1 9-inch baked pastry shell	**½ teaspoon salt**
½ pound Swiss cheese, grated	**⅛ teaspoon pepper**

1. Place bacon slices on 2 paper towels. Cover with another paper towel. Cook on BAKE for 5 to 6 minutes, or until almost crisp. Reserve bacon.
2. Combine onion and butter in a 1-quart casserole.
3. Cook, covered, on HIGH for 3 to 4 minutes, or until onion is limp.
4. Place cooked onion in prebaked pastry shell. Toss together cheese and flour and sprinkle over onion. Beat together eggs, milk, salt, and pepper. Pour over cheese.
5. Cook on BAKE for 7 to 8 minutes.
6. Place bacon strips on top of pie. Cook on BAKE for 4 to 5 minutes, or until custard is almost set. Let stand 10 minutes to finish cooking.

78

Pasta

Many recipes using combinations of sauces and pastas, rice, or noodles depend on having all of the separately heated ingredients ready to be mixed at a certain doneness. The microwave oven provides the perfect solution to this age-old problem.

In winter, when hot cereals provide the ideal breakfast, all members of the family can delight themselves with their own favorite without fuss or mess for the homemaker.

Since pasta, rice, and cereal expand when cooked, be sure to use a container of adequate size. If the pasta or rice is to be used in a casserole, it should be slightly firmer—thus cooked less—than if it is to be immediately eaten.

Pasta — How to Cook

1. Combine water with 1 tablespoon salad oil and 1 to 2 teaspoons salt in a 3-quart glass casserole or baking dish. Bring to a boil on HIGH. Water from the hot tap boils faster.
2. Add pasta. Stir. Cover with glass lid or plastic wrap. Cook on SIMMER until done. Pasta to be used in a casserole can be slightly firmer than pasta that is eaten at once.
3. Drain in a colander. Rinse with warm water. Serve.

Pasta	Covered Glass Container	Water	Minutes to Boil Water	Amount of Pasta	Minutes to Cook
Spaghetti	3-quart casserole or baking dish	4 cups	8 to 10 on HIGH	8 ounces	8 to 10 on SIMMER
Macaroni	3-quart casserole	3 cups	8 to 10 on HIGH	2 cups	10 to 12 on SIMMER
Egg noodles	3-quart casserole	6 cups	8 to 10 on HIGH	4 cups	10 to 12 on SIMMER
Lasagna noodles	3-quart, 13- by 9-inch baking dish	6 cups	10 to 12 on HIGH	8 ounces	14 to 15 on SIMMER

Rice — How to Cook

1. Select glass casserole or baking dish two to three times larger than amount of uncooked rice to be cooked.
2. Add salt and margarine to water according to rice package directions.
3. Bring seasoned water to a full boil on HIGH.
4. Stir in uncooked rice. Cover tightly with glass lid or plastic wrap. Cook on SIMMER.
5. Let cooked rice stand, covered, 5 minutes after being removed from oven.

Rice	Covered Glass Container	Water	Minutes to Boil Water	Amount of Rice	Minutes to Cook
Short-grain white rice	2 quart	2 cups	4 to 5 on HIGH	1 cup	13 to 15 on SIMMER
Long-grain rice	2 quart	2 cups	4 to 5 on HIGH	1 cup	15 to 17 on SIMMER
Wild rice	3 quart	3 cups	6 to 7 on HIGH	1 cup	35 to 40 on SIMMER
Brown rice	3 quart	3 cups	6 to 7 on HIGH	1 cup	40 on SIMMER
Quick-cooking white rice	1 quart	1 cup	3 to 4 on HIGH	1 cup	Let stand, covered, for 5 minutes, or until water is absorbed.

Cereal — How to Cook

How to cook quick-cooking or instant cereal.

1. Measure water, salt, and cereal into individual serving bowls. Stir lightly.
2. Follow package directions for amount of cereal to be used.
3. Cook on ROAST as directed in chart.
4. Stir after removing from oven. Let stand about 1 minute.

Servings	Water	Salt	Amount of Cereal	Minutes to Cook	Setting
1	¾ cup	¼ teaspoon	¼ or ⅓ cup, depending on cereal	2½ to 3	ROAST
2	¾ cup each	¼ teaspoon each	¼ to ⅓ cup each, depending on cereal	3½ to 4	ROAST

Green Noodles   6 servings

¼ cup butter or margarine
¼ cup all-purpose flour
1 teaspoon salt
¼ teaspoon hot-pepper sauce
2½ cups milk

1 cup diced sharp Cheddar cheese
¼ cup grated Parmesan cheese
3 cups cooked green noodles
3 hard-cooked eggs, halved

1. Put butter in a 1-quart measure.
2. Cook, covered, on HIGH for 30 seconds.
3. Remove and stir in flour, salt, and hot-pepper sauce to make a smooth paste.
4. Cook on HIGH for 30 seconds.
5. Gradually stir in milk. Cook, covered, on HIGH for 4 to 5 minutes, stirring once during cooking time.
6. Remove and stir briskly to make a smooth sauce. Add Cheddar cheese and Parmesan cheese and stir until cheese is melted.
7. Put noodles in a 1½-quart casserole. Add cheese sauce and mix lightly. Cook, covered, on BAKE for 7 to 8 minutes, or until piping hot.
8. Top with egg halves. Cook, covered, on BAKE for 3 minutes.
9. Let stand, covered, for 4 minutes before serving.

Noodles and Chicken   4 to 6 servings

1½ cups broken narrow egg noddles
2 to 3 cups cut-up cooked chicken or turkey
1 cup chicken stock
½ cup milk

½ teaspoon salt
⅛ teaspoon pepper
1 cup shredded Cheddar cheese
¼ cup sliced stuffed green olives

1. In a 2-quart casserole combine noodles, chicken, chicken stock, milk, salt, and pepper. Stir lightly.
2. Cook, covered, on HIGH for 8 to 10 minutes, stirring once, or until noodles are tender.
3. Stir in cheese and olives. Cook, covered, on LOW for 5 minutes, or until cheese is melted.

Following pages: Left, *Macaroni and Cheese (recipe, p. 84)*;
right, *Spaghetti Sauce (recipe, p. 85)*

Marcaroni and Cheese *(Illustrated on page 82)*   4 servings

2 tablespoons butter or
  margarine
2 tablespoons all-purpose
  flour
¼ teaspoon salt
½ teaspoon Worchestershire
  sauce
½ teaspoon prepared mustard

Freshly ground pepper to taste
1 cup milk
2 cups shredded Cheddar
  cheese, divided
1¼ cups uncooked macaroni,
  cooked*
¼ cup cracker crumbs
Tomato slices (optional)

1. Melt butter in a 4-cup measure on HIGH for 1 minute.
2. Stir in flour, salt, Worchestershire, mustard, and pepper. Gradually stir in milk. Cook, uncovered, on HIGH for 3 minutes. Stir. Cook on HIGH for 1 minute, or until smooth and thickened.
3. Stir in 1½ cups shredded cheese. Cook on ROAST for 30 seconds, or until cheese is melted.
4. Combine sauce and cooked macaroni. Pour into a 1½-quart casserole or baking dish. Combine remaining cheese and cracker crumbs. Sprinkle over top of casserole.
5. Cook, uncovered, on BAKE for 5 to 6 minutes, or until mixture is bubbling.
6. Let stand, covered, 5 minutes before serving. Top with tomato slices, if desired.

*Refer to pasta chart for cooking directions.

Macaroni Supreme   8 to 10 servings

1 pound lean ground beef
1 large onion, finely chopped
1 can (35 ounces) Italian-
  style tomatoes, undrained
1 package (10 ounces) frozen
  green peas
1 cup sliced fresh mushrooms
¾ cup dry red wine

¼ cup chopped parsley
1 teaspoon sugar
1½ teaspoons salt
Dash pepper
3 cups uncooked elbow
  macaroni, cooked,* divided
2 cups freshly grated
  Parmesan cheese, divided

1. Combine meat and onions in a 1-quart casserole or baking dish. Cook, covered, on HIGH for 3 minutes. Stir and cook on HIGH for 2 more minutes. Drain off excess grease.
2. Add tomatoes to ground beef. Crush whole tomato pieces with spoon. Add peas, mushrooms, wine, parsley, sugar, salt, and pepper to casserole.
3. Cook, covered, on SIMMER for 30 minutes.
4. In a 13- by 9-inch glass baking dish, layer half the sauce, half the cooked macaroni, and 1 cup cheese. Add remaining macaroni and sauce. Top with remaining cheese.
5. Cook, covered, on BAKE for 10 to 15 minutes, or until cheese melts and casserole is bubbling.
6. Let stand 5 minutes before serving.

*Refer to pasta chart for cooking directions.

Spaghetti Sauce *(Illustrated on page 83)*   about 2 quarts

½ pound ground beef	2 teaspoons salt
½ cup chopped onions	2 teaspoons oregano
2 cloves garlic, minced	¼ teaspoon basil
1 can (28 ounces) tomatoes	¼ teaspoon ground thyme
2 cans (6 ounces each) tomato paste	Freshly ground pepper to taste

1. Crumble beef into a 3-quart casserole. Add onion and garlic.
2. Cook, uncovered, on HIGH for 5 minutes, stirring once to break up meat.
3. Add remaining ingredients. Break up whole tomatoes with a fork or potato masher.
4. Cook, covered, on SIMMER for 25 to 30 minutes, or until mixture is well blended and slightly thickened.
5. Cover and let stand about 5 minutes.
6. Serve over hot cooked spaghetti.

Marinara Sauce   2 to 3 servings

1 can (2 pounds 3 ounces) peeled Italian plum tomatoes	2 teaspoons dried basil
½ cup salad oil	1 tablespoon chopped parsley
¼ cup thinly sliced garlic	Salt and freshly ground pepper to taste

1. Empty tomatoes into a bowl. Crush with hands so that there are no whole tomatoes left.
2. Put oil and garlic in a 1½-quart casserole. Cook at HIGH for 3 minutes.
3. Add basil, parsley, and tomatoes. Cover and cook on REHEAT for 10 minutes. Cook on BAKE for 5 minutes.
4. Let stand 5 minutes. Season to taste with salt and pepper and serve over hot cooked spaghetti.

Macaroni Goulash   6 to 8 servings

1 cup uncooked macaroni	½ teaspoon basil
1 pound ground beef	1 teaspoon salt
1 can (1 pound) tomato puree, or 1 can (1 pound) tomatoes packed in puree	Pepper to taste
½ teaspoon sugar	1 tablespoon chopped parsley

1. Cook macaroni on conventional range according to package directions. Set aside.
2. Crumble beef into a 3-quart casserole. Cook, uncovered, on HIGH for 4 to 5 minutes, stirring once to break up meat.
3. Add puree, sugar, basil, salt, pepper to taste, and parsley. Stir in macaroni.
4. Cook, covered, on HIGH for 8 minutes, stirring once.
5. Cook, covered, on LOW for 4 to 5 minutes before serving.

Variation: To stretch amount, use 1 can (1 pound 12 ounces) tomato puree and increase macaroni to 1½ cups.

Manicotti

8 to 10 servings

1 package (8 ounces) manicotti
  noodles
1 package (1 pound) ricotta cheese
½ pound mozzarella cheese, grated
Parmesan cheese
3 tablespoons chopped parsley,
  divided
3 teaspoons sugar, divided
1 egg, lightly beaten

Salt and pepper to taste
2 sweet Italian sausages
1 clove garlic, minced
1 medium onion, minced
1 pound ground beef
1 can (12 ounces) tomatoes, mashed
1 can (16 ounces) tomato sauce
½ teaspoon basil

1. Cook manicotti noodles according to package directions for 12 minutes on conventional range. Drain and reserve.
2. Combine ricotta, mozzarella, 3 tablespoons Parmesan cheese, 1 tablespoon parsley, 2 teaspoons sugar, egg, and salt and pepper to taste. Blend well and reserve.
3. Remove sausage from casings. Crumble into a 3-quart casserole. Add garlic, onion, and 2 tablespoons chopped parsley.
4. Cook, covered, on HIGH for 3 minutes, stirring once.
5. Crumble ground beef on top of sausage meat and toss lightly. Cook, covered, on HIGH for 5 minutes, stirring once to break up meat during cooking period.
6. Add tomatoes, tomato sauce, basil, 1 teaspoon sugar, and salt and pepper to taste.
7. Cook, covered, on HIGH, stirring once.
8. Pour a thin layer of meat sauce on the bottom of two 2-quart baking dishes or flat casseroles.
9. Fill cooked manicotti tubes with cheese mixture. Place 10 filled tubes close together in each casserole. Cover with remaining meat sauce.
10. Cook, covered tightly, on HIGH for 20 minutes, or until hot.
11. Remove cover, and sprinkle ¼ cup grated Parmesan on top of casserole. Cook on ROAST for 2 minutes, or just until cheese is melted.

Note: This recipe makes 2 casseroles. Use one immediately and freeze the second for later use.

Vegetables

Because vegetables are cooked in very little water, in a tightly covered dish, and for a short length of time, they retain more nutrients and, most important, keep their color and flavor better when cooked in a microwave oven. Vegetables should be crisp-cooked, almost Chinese style, because—as most foods do—they will continue to cook after they are removed from the oven.

In order to assure even cooking, vegetables should be cut in uniform pieces and stirred during the cooking time. All fresh vegetables should be cooked on HIGH to capture their ultimate goodness. Parts that take longer should always be placed at the outside of the dish. Whole, unpeeled vegetables should always be pierced or pricked to allow steam to escape.

Fresh and Frozen Vegetables — How to Cook

Microwave all fresh and frozen vegetables on HIGH.

Vegetable	Amount	Approximate Cooking Time in Minutes	Preparation
Artichokes (about 3⅓ inches in diameter)	1 2 4 frozen hearts, 10-ounce package	4 to 5 8 to 9 12 to 15 5 to 6	Wash thoroughly. Cut tops off each leaf with sharp scissors. Cut 1 inch off top with a sharp knife.
Asparagus	fresh, 1 pound frozen, 10-ounce package	6 to 7 7 to 8	Wash thoroughly to remove sand. Snap off tough base and discard.
Beans: green and wax	fresh, 1 pound frozen, 9-ounce package	12 to 14 7 to 8	Remove ends. Wash well and leave whole or break in pieces.
Beets	4 medium	18 to 20	Scrub beets. Leave 1 inch of top on beet. Peel and cut, or slice when beet is cooked.
Broccoli	fresh, 1 to 1½ pounds frozen, 10-ounce package	8 to 9 8 to 10	Remove tough outer leaves. Leave in stalks or cut stem in pieces and flowers in pieces.
Brussels sprouts	fresh, 1 pound frozen, 10-ounce package	8 to 9 8 to 9	Remove outside leaves if wilted. Cut off stems. Wash.
Cabbage	½ medium head	5 to 6	Remove outside wilted leaves. Shred cabbage for faster cooking.
Carrots	4 carrots, cut 6 carrots, cut frozen, 10-ounce package	5 to 6 6 to 7 7 to 8	Peel and cut off tops. Slice, dice, or cut in slivers. Tiny baby carrots are good when left whole.
Cauliflower	1 in flowerets 1 whole frozen, 10-ounce package	8 to 9 8 to 10 7 to 8	Wash and remove outside leaves. Leave whole or break into several flowerets.
Celery	6 stalks	8 to 9	Clean stalks thoroughly. Use outer branches for cooking; reserve inner stalks for eating raw. Slice crosswise into half-moons.
Corn, kernel	frozen, 10-ounce package	5 to 6	
Corn on cob	4 ears	6 to 7	Husk and wrap each ear in a square of waxed paper. Place on glass bottom in oven, cooking no more than 4 at one time.
	4 ears frozen, 2 ears frozen, 4 ears	8 6 to 8 10 to 12	*Or* remove husks from corn. Place in a square baking dish. Cover with a square of waxed paper.
Eggplant	1 medium	5 to 6	Wash and peel. Cut into slices or cubes.
Okra	½ pound frozen, 10-ounce package	3 to 5 7 to 9	Wash thoroughly. Leave whole or cut in thick slices.

Vegetable	Amount	Approximate Cooking Time in Minutes	Preparation
Parsnips	4 medium	8 to 9	Peel and cut in quarters.
Peas	fresh: 1 pound	7 to 8	Shell peas. Rinse well.
	fresh: 2 pounds	8 to 10	
	frozen, 10-ounce package	5 to 6	
	pea pods, frozen, 6-ounce package	3 to 5	
Potatoes, sweet or yam	2	6 to 7	Wash and scrub well. Pierce with a fork all over. Place on paper towels, 1 inch apart.
	4	8 to 11	
	6	10 to 12	
Potatoes, white	1	4 to 4½	Wash and scrub well. Pierce with a fork all over. Place on paper towels, about 1 inch apart.
	2	7 to 9	
	4	10 to 13	
	4	12 to 16	To boil: Peel potatoes and cut in quarters. Place in a glass casserole. Add water, cover tightly, and cook.
Spinach	fresh, 1 pound	6	Wash well. Remove tough stems or any wilted leaves. Drain as much water off as possible.
	frozen, 10-ounce package	7 to 8	
Squash, acorn or butternut	1 whole	10 to 12	Scrub and leave whole. Pierce with a knife in several spots. Cut and remove seeds to serve when cooked.
	cut, halves	8 to 10	Squash may be halved and placed, cut side down, on paper towels. Reverse for last minutes of cooking.
Squash, zucchini	2	7 to 8	Wash, do not peel. Cut in slices.
Turnips or rutabaga	4 small	13 to 15	Peel. Wash and cut in cubes.
Vegetables, mixed	frozen, 10-ounce package	7 to 8	

Asparagus Vinaigrette Bundles
5 to 6 servings

2 dozen fresh asparagus spears
6 tablespoons cooking oil
3 tablespoons vinegar
⅛ teaspoon hot-pepper sauce
½ teaspoon sugar
¼ teaspoon salt
1 small onion, sliced
Pimiento strips

1. Cook asparagus spears 6 to 7 minutes. Cool. Place in a shallow dish.
2. Combine remaining ingredients, except pimiento, and blend well. Pour over asparagus and let stand in refrigerator several hours or overnight.
3. Wrap pimiento strips around 4 or 5 asparagus spears. Serve as a cold vegetable or place on lettuce cups and serve as a salad.

Harvard Beets

4 servings

1 can (1 pound) diced or sliced beets
¼ cup sugar
1 tablespoon cornstarch

½ teaspoon salt
Freshly ground pepper to taste
¼ cup vinegar

1. Drain beets, reserving liquid. Pour beet liquid into a 1-cup measure and add enough water to make 1 cup of liquid.
2. Combine sugar, cornstarch, salt, pepper, and vinegar in a 1-quart casserole or bowl. Stir in beet liquid.
3. Cook, uncovered, on HIGH for 2½ to 3 minutes, stirring occasionally, until mixture thickens and is clear.
4. Add beets and stir lightly.
5. Cook, covered, on HIGH for about 3 minutes, or until beets are piping hot.

Pickled Beets

2 cups

1 can (1 pound) sliced beets
⅓ cup sugar

⅓ cup vinegar
1 teaspoon pickling spice

1. Drain beets, reserving ⅓ cup of the beet liquid. Place beets in a 1-quart casserole with sugar, beet liquid, and vinegar. Tie pickling spice in a small square of cheesecloth and add to beets.
2. Cook, covered, on HIGH for 4 to 5 minutes, or until mixture comes to a boil.
3. Cool and remove bag of spices. Refrigerate up to 2 weeks.

Beets in Orange Sauce

4 servings

1 can (1 pound) diced beets
1 tablespoon cornstarch
¾ teaspoon salt
1½ tablespoons sugar

¼ cup orange juice
2 tablespoons lemon juice
½ teaspoon grated orange peel
1 tablespoon butter or margarine

1. Drain beets, reserving liquid. Pour liquid into a measuring cup and add enough water to make ½ cup liquid.
2. Combine cornstarch, salt, sugar, and orange juice in a 1-quart bowl. Stir in beet liquid.
3. Cook, uncovered, on HIGH for 2½ to 3 minutes, or until mixture comes to a boil and is clear.
4. Add lemon juice, orange peel, and butter. Stir to melt butter. Add beets.
5. Cook, uncovered, on HIGH for about 3 minutes, or until beets are piping hot.

Peppered Beans

3 to 4 servings

1 pound green beans
2 tablespoons olive oil
½ sweet red or green pepper, seeded and cut in slivers

¼ cup blanched slivered almonds
Salt and pepper to taste

1. Cook beans according to preceding directions. Cover and let stand.
2. Combine oil, red or green pepper, and almonds in a 1-quart casserole.
3. Cook, uncovered, on HIGH for 3 to 4 minutes, or until peppers are limp.
4. Toss with green beans. Season to taste with salt and pepper.

Green Beans Italian

6 servings

2 packages (10 ounces each) frozen green beans
1 small onion, thinly sliced

¾ cup bottled Italian dressing
3 strips cooked bacon

1. Place green beans in a 1½-quart casserole or saucepan, icy side up.
2. Cook, covered, on HIGH for 7 to 8 minutes, or until almost tender, stirring once during cooking time.
3. Add onion and Italian dressing.
4. Cook, covered, on HIGH for 3 to 4 minutes, or until beans are crisply tender.
5. Serve hot, topped with crumbled cooked bacon.

Green Beans Piquant

3 to 4 servings

1 pound green beans
2 tablespoons butter or margarine
1 teaspoon prepared mustard

1 tablespoon Worcestershire sauce
Salt and pepper to taste

1. Cook beans according to directions on page 95.
2. When beans are tender, add remaining ingredients and toss lightly until butter is melted and beans are well coated with mixture.

Savory Green Beans

3 to 4 servings

1 pound green beans
2 tablespoons olive oil
1 clove garlic, minced
1 teaspoon catsup

1 teaspoon Worcestershire sauce
¼ teaspoon savory
Salt to taste

1. Cook green beans according to directions. Cover and keep warm.
2. Combine olive oil and garlic in a 1-quart serving bowl.
3. Cook, uncovered, on HIGH for 1½ minutes, or until garlic is tender.
4. Add catsup, Worcestershire, and savory. Add hot beans and toss lightly. Season to taste with salt.

Lima Beans Parmesan

3 to 4 servings

1 package (10 ounces) frozen baby lima beans
¼ cup chicken bouillon
1 bay leaf

1 clove garlic
Salt and pepper to taste
Grated Parmesan cheese

1. Place lima beans in a 1½-quart casserole. Add chicken bouillon, bay leaf, and garlic.
2. Cook, covered, on HIGH for 9 to 10 minutes.
3. Remove bay leaf and garlic. Season to taste with salt and pepper. Serve with Parmesan cheese sprinkled on top of beans.

Broccoli Indienne

3 to 4 servings

1 bunch broccoli, about 1½ pounds
⅓ cup chicken bouillon
1 bay leaf
¼ teaspoon thyme

2 tablespoons lemon juice
1 tablespoon butter or margarine
Salt to taste

1. Wash broccoli. Cut off tough bottom stalks and discard. Remove leaves and peel tough skin from stalks. Cut stems in chunks, leaving flowerets intact.
2. Place stem chunks on sides and flowerets in center of a 1½-quart casserole. Add chicken bouillon, bay leaf, and thyme.
3. Cook, covered, on HIGH for 5 to 7 minutes, or just until tender.
4. Discard bay leaf and add lemon juice and butter. Toss lightly and season to taste with salt.

Norwegian Cabbage

3 to 4 servings

½ head cabbage
½ cup dairy sour cream

½ teaspoon caraway seeds
Salt and pepper to taste

1. Shred cabbage. Cook on HIGH for 6 to 7 minutes, or until tender. Drain.
2. Toss lightly with sour cream and caraway seeds. Season to taste with salt and pepper. Serve hot.

Sweet and Sour Cabbage

4 to 6 servings

4 cups shredded cabbage
2 apples, peeled, cored, and finely chopped
¼ cup brown sugar

1 teaspoon salt
⅛ teaspoon pepper
½ cup butter
¼ cup vinegar

1. Place cabbage in a 1½-quart casserole. Combine remaining ingredients. Pour over cabbage.
2. Cook, covered, on HIGH for 6 to 7 minutes, or until cabbage is tender. Stir once during cooking period.

Sweet and Sour Celery

3 to 4 servings

2 cups thinly sliced celery
1 bay leaf
3 whole cloves
2 tablespoons sugar

3 tablespoons vinegar
2 tablespoons butter or margarine
Salt and pepper to taste

1. Place celery, bay leaf, cloves, and ¼ cup water in a 1½-quart casserole.
2. Cook, covered, on HIGH for 8 to 9 minutes, or just until celery is crisply tender.
3. Add sugar, vinegar, and butter. Toss lightly. Cook, covered, on HIGH for 1 minute, or until butter is melted and celery is hot. Season to taste.

Cauliflower and Tomatoes

4 servings

1 medium head cauliflower	½ teaspoon salt
1 clove garlic	½ cup cooked tomatoes
3 tablespoons olive oil	2 tablespoons grated Parmesan cheese

1. Remove outer leaves and stalks from cauliflower. Separate into flowerets. Cook according to directions but undercook just slightly.
2. Combine garlic and olive oil in a 1½-quart casserole. Cook, uncovered, on HIGH for 1 minute.
3. Remove garlic. Add hot drained cauliflower to oil.
4. Cook, uncovered, on HIGH for 2 minutes.
5. Add salt and tomatoes. Cook, covered, on HIGH for 4 minutes, or until piping hot. Stir once during cooking time.
6. Sprinkle top with cheese and serve immediately.

Cheesed Cauliflower

4 servings

1 medium head cauliflower	Dash of pepper
3 tablespoons olive oil	¼ cup fine dry bread crumbs
1 large onion, thinly sliced	¼ cup grated Cheddar cheese
¼ teaspoon salt	

1. Remove outer leaves and stalks from cauliflower. Separate into flowerets. Cook according to preceding directions.
2. Put olive oil in a 1-quart measure. Add onion. Cook, uncovered, on SAUTE for about 4 minutes, or until onions are limp. Stir once during cooking time.
3. Add salt, pepper, and bread crumbs.
4. Drain cauliflower and leave in original casserole. Top with hot onion mixture. Sprinkle cheese over top.
5. Cook, uncovered, on ROAST for 1½ to 2 minutes, or until cheese is melted and cauliflower is piping hot.

Artichoke Hearts with Mushrooms

3 to 4 servings

1 package (10 ounces) frozen artichoke hearts	½ teaspoon lemon juice
1 can (4 ounces) sliced mushrooms	Salt and pepper
1½ teaspoons cornstarch	Onion salt
2 tablespoons dry sherry	Garlic salt
2 tablespoons butter or margarine	1 tablespoon chopped parsley

1. Cook artichoke hearts according to preceding directions. Set aside.
2. Drain mushrooms, reserving liquid.
3. Combine cornstarch, sherry, butter, lemon juice, and mushroom liquid in a 1-quart bowl.
4. Cook, uncovered, on HIGH for 30 seconds to 1 minute, or until mixture is thickened and clear. Blend well.
5. Season to taste with salt, pepper, onion salt, and garlic salt. Stir in parsley. Cut artichoke hearts in half and add them, along with the mushrooms. Stir gently.
6. Heat, covered, on HIGH for 2 to 3 minutes, or until mixture is piping hot.

Cauliflower and Tomatoes

Stuffed Eggplant

4 servings

2 medium eggplants
2 medium onions, chopped
1 pound ground lamb
1 beef bouillon cube
1 can (8 ounces) tomato sauce

½ teaspoon oregano
2 tablespoons chopped parsley
½ teaspoon salt
¼ teaspoon pepper
½ cup dry bread crumbs

1. Wash eggplant and cut in half lengthwise. Scoop out insides, leaving a shell 1 inch thick. Chop eggplant pulp in medium chunks. Set aside.
2. Put onion in a 1½-quart casserole. Crumble in lamb. Cook, covered, on HIGH for about 5 minutes, or just until lamb loses pink color. Drain fat.
3. Dissolve bouillon cube in ½ cup hot water. Stir into cooked lamb with 3 tablespoons tomato sauce and the chopped eggplant pulp.
4. Cook, covered, on HIGH for about 5 minutes, stirring occasionally.
5. Remove from oven; stir in oregano, parsley, salt, and pepper. Fill eggplant halves with mixture. Sprinkle bread crumbs over top. Streak remaining tomato sauce over top of crumbs.
6. Place eggplant halves in a glass baking dish. Cook, covered, on REHEAT for 8 minutes, or just until eggplant is tender.

Stuffed Green Peppers

4 to 6 servings

4 large green peppers
1 pound ground beef
1 medium onion, finely chopped
1 teaspoon salt

¼ teaspoon pepper
1½ cups cooked rice
1 can (16 ounces) tomato sauce

1. Wash peppers. Cut in half lengthwise and remove seeds and white membrane.
2. Crumble beef into a 1½-quart casserole. Add onion. Cook, uncovered, on HIGH for about 5 minutes, stirring once during cooking period. Cook until meat loses its red color.
3. Stir in salt, pepper, rice, and half of the tomato sauce. Fill green pepper halves with mixture, mounding mixture on top. Place in a glass baking dish. Top each pepper with a dribble of remaining tomato sauce.
4. Cook, covered, on HIGH for 8 to 10 minutes, or just until peppers are tender.

Eggs in Nests

8 servings

8 small, firm-ripe tomatoes
¼ cup chopped parsley
¼ cup butter or margarine

1 large onion, chopped
8 eggs
Salt and pepper to taste

1. Cut tops from tomatoes. Scoop out pulp and turn shells upside down on paper towels to drain. Discard seeds, chop pulp, and mix with parsley.
2. Combine butter and onion in a small mixing bowl. Cook, covered, on HIGH for about 4 minutes. Add parsley-tomato pulp mixture.
3. Stir mixture well and divide into tomato shells. Break 1 egg into each tomato shell. Season lightly with salt and pepper.
4. Place tomatoes in a baking dish. Cook, covered, on BAKE for 1 to 1½ minutes, or until eggs are set to desired degree of doneness.

Cranberry Carrots

4 servings

6 to 8 carrots
¼ cup butter or margarine

¼ cup jellied cranberry sauce
Salt and pepper to taste

1. Cook carrots according to directions. Add about 1 minute to cooking time for the larger number of carrots.
2. Place butter in a 1½- to 2-quart casserole. Cook, covered, on HIGH for 1½ minutes, or until butter is melted.
3. Add cranberry sauce. Cook, covered, on HIGH for 1 minute. Stir and cook until cranberry sauce is melted.
4. Add cooked carrots and toss gently. Season to taste with salt and pepper.

Tangy Glazed Carrots

3 to 4 servings

6 carrots
⅓ cup orange juice
2 tablespoons sugar
¼ teaspoon ground cloves

¼ teaspoon salt
½ jar (5 ounces) pineapple cheese spread

1. Peel carrots. Slice into rounds. Cook according to directions.
2. Combine juice, sugar, cloves, salt, and cheese spread. Blend thoroughly. Pour mixture over hot cooked carrots.
3. Cook, uncovered, on HIGH for about 1½ minutes, or until cheese melts and mixture is piping hot.

Corn Pudding

4 servings

2 tablespoons butter or margarine
2 tablespoons all-purpose flour
1 can (1 pound) whole kernel corn, drained

2 cups milk
2 eggs, well beaten
1 teaspoon salt
½ teaspoon pepper

1. Melt butter in a 1½-quart casserole on ROAST for 30 seconds.
2. Stir in flour to make a smooth paste. Add remaining ingredients and blend well.
3. Cook, covered, on HIGH for 9 minutes. Stir once during cooking period.
4. Cook, covered, on WARM for 3 minutes before serving.

Note: Good with ham or pork chops for a hearty winter dinner.

Saucy Eggplant

6 servings

4 slices bacon
1 medium onion, chopped
1 medium green pepper, seeded and chopped
1 1-pound eggplant

2 teaspoons salt
¼ teaspoon pepper
1 can (8 ounces) tomato sauce
½ cup grated Parmesan cheese

1. Place bacon on inverted saucer in an oblong glass baking dish. Cover with waxed paper. Cook on HIGH for about 5 minutes, or until bacon is crisp. Reserve bacon and drippings.
2. Put onion and pepper in a 1½-quart casserole. Pour bacon fat over top of onion.
3. Cook, covered, on HIGH for 4 minutes.
4. Peel eggplant and cut in cubes. Add to onion mixture in casserole with salt, pepper, tomato sauce, and 1 cup water.
5. Cook, covered, on HIGH for 6 minutes.
6. Remove from oven and sprinkle cheese over top of mixture. Crumble bacon and sprinkle over top of cheese.
7. Cook, covered, on HIGH for 6 minutes.
8. Cook, covered, on WARM for 4 minutes before serving.

Sautéed Green Peppers

4 to 6 servings

4 medium green peppers
1 tablespoon olive oil
1 tablespoon butter

Garlic salt
Pepper

1. Remove stem ends and seeds from peppers. Rinse well and cut into ¼-inch-thick strips.
2. Heat oil and butter in a 2-quart glass baking dish on HIGH for 1 minute. Add pepper strips. Cook, covered, on HIGH for 2½ minutes.
3. Season to taste with garlic salt and pepper. Toss lightly. Cook, uncovered, on HIGH for 1 to 2 minutes, or until peppers are crisply tender.

Sautéed Mushrooms

2 to 4 servings

½ pound fresh mushrooms
1 clove garlic, minced

⅓ cup butter or margarine

1. Clean mushrooms and slice. Put in an 8-inch round dish or a skillet. Add garlic and butter.
2. Cook, covered, on SAUTE for 4 to 5 minutes.
3. Serve with roast beef or steak, or on crisp toast as a main dish.

Hashed Potatoes

4 servings

⅓ cup butter or margarine
½ cup coarsely chopped onions

4 baked potatoes, cold
Salt and pepper to taste

1. Put butter and onion in a 1½-quart casserole. Cook, uncovered, on ROAST for 4 to 5 minutes, stirring occasionally.
2. Peel potatoes and cut in small chunks. Stir into onion in casserole. Season to taste. Cook, uncovered, on HIGH for 4 to 5 minutes, stirring occasionally.

Scalloped Potatoes

6 to 8 servings

5 cups (about 6 medium) peeled and thinly sliced raw potatoes
3 onions, sliced
Salt and pepper to taste
3 teaspoons dry mustard, divided
3 tablespoons grated Parmesan cheese, divided

3 tablespoons all-purpose flour, divided
3 tablespoons butter or margarine
3 cups milk
Paprika

1. Line bottom of a 3-quart casserole with one-third of the potatoes and cover with one-third of the onion. Add salt and pepper to taste, sprinkle on 1 teaspoon mustard, 1 tablespoon cheese, and 1 tablespoon flour. Repeat process twice with remaining potatoes, onion, seasonings, cheese, and flour. Dot with butter. Pour milk over top. Sprinkle with paprika.
2. Cook, covered, on HIGH for 20 minutes.
3. Cook, uncovered, on ROAST for 15 minutes, or until potatoes are tender. Let stand for 5 minutes before serving.

Candied Sweet Potatoes

6 servings

6 medium sweet potatoes
1 cup brown sugar, firmly packed

2 tablespoons butter or margarine

1. Cook sweet potatoes according to preceding directions. Peel and slice. Arrange in a 2-quart casserole.
2. Combine sugar, butter, and ⅓ cup water in a 1-quart measure. Cook, uncovered, on ROAST for 3 to 4 minutes, or until mixture is well blended and hot.
3. Pour over top of potatoes. Cook, covered, on HIGH for 7 to 8 minutes, or until heated through, spooning glaze over potatoes occasionally.

Sandwiches

In the microwave oven, sandwiches heat very quickly. They need not be hot, only warmed, as they will continue to cook while they stand. When meat is used to make sandwiches, remember that thin-sliced meat heats faster than thicker slices. A meat and cheese sandwich of fairly generous proportions takes only fifty seconds as a general rule; this insures that it is not so hot that you have to wait for it to cool to eat it. Franks can be cooked alone or in the bun. Always heat sandwiches and breads on paper plates, napkins, or paper towels to absorb the steam, which can make the bread soggy.

Open-faced or closed—meat, fish, or cheese—sandwiches are a snap using the microwave oven.

Brapples
4 servings

2 to 3 medium apples	½ cup chopped walnuts
¼ teaspoon lemon juice	4 slices white bread, buttered
¾ cup brown sugar	4 slices process American cheese

1. Peel, core, and slice apples in very thin slices.
2. Mix with lemon juice, brown sugar, and walnuts.
3. Divide over bread slices. Top with cheese.
4. Place in a 9-inch square dish.
5. Cook, uncovered, on HIGH for 2 to 3 minutes, or until apples are tender.

Variation: Turn these into a last-minute dessert by topping each with a table-spoon of sour cream and a sprinkling of brown sugar and nuts.

Bermuda Grill
6 servings

2 cups chopped Bermuda onions	½ cup Sauterne
1 teaspoon salt	12 slices Swiss cheese
¼ teaspoon white pepper	12 slices rye bread, toasted

1. Place chopped onion in a shallow dish; sprinkle with salt and pepper. Add Sauterne.
2. Cover and let stand at least 1 hour, stirring every 15 minutes; drain.
3. Place a slice of cheese on each of 6 toast slices.
4. Divide marinated onion over sandwiches and top with another slice of cheese, then a second slice of bread.
5. Place each sandwich on a paper plate. Cook one at a time on REHEAT for 1 minute, or until cheese is melted.

Hot Salad Cheesewiches
4 servings

1 cup shredded Cheddar cheese	⅛ teaspoon chili powder
½ cup diced cucumber	4 slices bread
1 tablespoon minced onion	4 large thick tomato slices
¼ cup dairy sour cream	8 slices dill pickle
⅛ teaspoon pepper	Paprika

1. Combine cheese, cucumber, onion, sour cream, and seasonings.
2. Toast bread lightly. Place each slice of bread on a paper plate.
3. Arrange a tomato slice and 2 pickle slices on each slice of bread.
4. Divide the cheese mixture over the 4 slices, and sprinkle with paprika.
5. Cook each sandwich separately on REHEAT for 1 to 1½ minutes, or until cheese melts and mixture is thoroughly heated.

Baked Brunchwiches

6 servings

12 to 14 slices white bread	6 eggs
Butter or margarine	2 cups milk
½ pound grated Cheddar cheese	½ teaspoon salt
1 teaspoon dry mustard	Paprika

1. Remove crust from bread slices. Butter each slice and cut in quarters or small chunks. Put a layer of bread in a 2- to 3-quart casserole. Sprinkle half the cheese over bread. Sprinkle half the mustard over cheese. Repeat layers with bread, cheese, and mustard.
2. Beat eggs with milk and salt. Pour over top of bread and cheese layers. Sprinkle generously with paprika.
3. Cover with foil or casserole lid and refrigerate overnight.
4. Cook on BAKE for 15 minutes.
5. Cook on HIGH for 10 minutes.
6. Remove from oven and cover with foil. Let stand 3 to 4 minutes before serving.

Cheese Roll-Ups

6 servings

4 slices bacon	¼ teaspoon Worcestershire sauce
1 loaf (1 pound) unsliced white bread	1 can (10½ ounces) condensed cream
1 cup grated process American cheese	of mushroom soup, undiluted
¼ cup chopped stuffed olives	¼ cup milk

1. Cook bacon according to directions on page 60 and reserve.
2. Cut crusts from loaf and cut loaf into 6 horizontal slices.
3. Crumble bacon and combine with cheese, olives, Worcestershire, and ⅓ cup of the mushroom soup.
4. Spread 3 tablespoons of the mixture on each slice of bread. Roll up, jelly-roll fashion.
5. Place each roll-up on a paper plate. Cook one at a time on HIGH for 1½ minutes, or until heated through.
6. Combine remaining mushroom soup and milk in a 2-cup glass measuring cup. Cook on BAKE for 4 to 4½ minutes, or until piping hot.
7. Serve mushroom sauce over roll-ups.

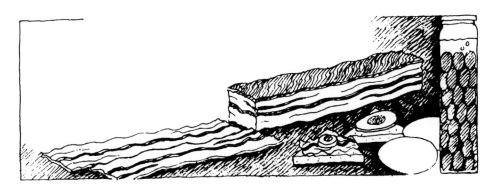

Reuben Sandwich

4 servings

8 slices dark rye or pumpernickel bread
Butter or margarine
½ pound thinly sliced corned beef

1 can (8 ounces) sauerkraut, drained
Thousand Island dressing
4 slices Swiss cheese

1. Toast bread. Butter lightly.
2. Arrange sliced corned beef on 4 slices of toast. Divide sauerkraut among sandwiches. Top with Thousand Island dressing. Top each with a slice of Swiss cheese. Top with other slices of toast, buttered side down.
3. Place each sandwich on a paper plate.
4. Cook one at a time on REHEAT for about 1 minute, or just until cheese is melted.

Hot Swiss Tuna

4 servings

4 hamburger buns
1 can (6½ or 7 ounces) tuna fish, drained and flaked
½ cup finely shredded Swiss cheese
1 cup chopped celery

¼ cup mayonnaise
2 tablespoons catsup
1 teaspoon lemon juice
Salt and pepper to taste

1. Split hamburger buns.
2. Combine tuna, cheese, celery, mayonnaise, catsup, and lemon juice. Season to taste with salt and pepper.
3. Divide tuna mixture among four buns.
4. Wrap each sandwich in waxed paper and twist ends of paper.
5. Cook one at a time on REHEAT for 1 to 1½ minutes, or until rolls are hot.

Barbecued Beef Strips on Buns

6 servings

½ cup butter or margarine
1 pound top round steak, cut into thin strips
1 teaspoon salt
⅛ teaspoon garlic salt
⅛ teaspoon pepper
1 can (10½ ounces) beef broth

1 can (8 ounces) tomato paste
2 tablespoons cornstarch
1 teaspoon sugar
1 can (4 ounces) sliced mushrooms, drained
¼ cup dry red wine
6 buns

1. Put butter in a 2- or 3-quart casserole. Cook, uncovered, on HIGH for 1 minute, or just until butter melts.
2. Put strips of meat in casserole and toss lightly so that meat is coated with butter. Add salt, garlic salt, and pepper.
3. Cook, uncovered, on HIGH for 5 to 6 minutes, stirring once.
4. Add beef broth and tomato paste. Combine cornstarch and sugar and stir into mixture.
5. Cook, covered, on HIGH for 4 minutes, stirring once.
6. Add mushrooms and wine. Cook, covered, on HIGH for 1 to 2 minutes, or until mushrooms are hot.
7. Let stand, covered, on LOW for 4 minutes before serving.
8. Serve beef strips, with sauce, in heated buns.

Reuben Sandwich

Ranchburgers

6 servings

6 slices bacon

1½ cups grated process American
cheese

2 tablespoons finely chopped onion

¼ cup catsup

1 tablespoon prepared mustard

6 sandwich buns, split

1. Cook bacon according to directions on page 60. Crumble bacon and combine with remaining ingredients except buns.
2. Spread 3 tablespoons cheese mixture on bottom half of each bun and cover
with bun top.
3. Wrap each sandwich in waxed paper and twist ends of paper.
4. Cook each sandwich separately on REHEAT for 1 to 1½ minutes, or until
heated through.

Second-Act Ham

3 to 4 servings

¾ cup hollandaise sauce

4 slices bread, toasted

Ham slices from leftover baked ham

1 can (8 ounces) asparagus, drained,
or 9 to 12 fresh asparagus, cooked

2 hard-cooked eggs, sliced (optional)

1. Prepare hollandaise sauce, using your favorite recipe. Cover with waxed
paper and set aside.
2. Arrange toast, cut in quarters, in individual casseroles. Put desired amount of
ham on top of toast. Arrange 3 or 4 asparagus spears on top of ham. Add ½
egg, sliced, on top if desired.
3. Cook, covered, 1 casserole at a time, on BAKE for about 2 minutes, or until
heated through.
4. Top with hollandaise sauce. Cover with waxed paper and heat on BAKE for 1
minute. If hollandaise sauce is cold, double the heating time.

Sardine Buns

6 servings

¼ pound process American cheese,
cubed

5 hard-cooked eggs, chopped

½ cup drained mashed sardines

1 tablespoon minced green pepper

2 tablespoons minced onion

3 tablespoons chopped stuffed olives

2 tablespoons pickle relish, drained

½ cup mayonnaise

6 hamburger buns, split and buttered

1. Combine all ingredients except buns.
2. Fill each bun with cheese mixture.
3. Wrap each sandwich in waxed paper and twist ends of paper.
4. Cook, one at a time, on REHEAT for 1 to 1½ minutes, or until rolls are hot.

Note: These may be prepared in advance and refrigerated in their waxed paper
wrapping. Cook before serving.

106

Open-Face Hamburgers

6 servings

1 pound ground beef
1 teaspoon salt
1 teaspoon oregano
½ teaspoon dry mustard
Freshly ground pepper to taste
1 tablespoon instant minced onion

½ cup tomato juice
1 cup shredded Cheddar cheese
3 hamburger buns, halved and toasted
6 slices tomato
2 tablespoons butter or margarine

1. Crumble ground beef into a 1½-quart casserole or an 8-inch round glass cake dish.
2. Cook, uncovered, on HIGH for 5 to 6 minutes, or until meat has lost its red color. Stir once during cooking time.
3. Break up meat with a fork. Add salt, oregano, dry mustard, pepper, onion, tomato juice, and cheese. Stir thoroughly.
4. Cook, covered, on BAKE for 2 to 3 minutes, or until cheese is melted and mixture is piping hot.
5. Place toasted hamburger halves on a broiler pan. Spoon hamburger mixture on top of each bun. Top each with a slice of tomato. Dot with butter. Place under broiler on conventional range and broil until tomatoes are lightly browned.

Old Mystic

6 servings

1 can (7½ ounces) crab meat, drained
1 can (5 ounces) shrimp, drained
2 packages (3 ounces each) cream cheese, at room temperature
½ cup chopped almonds
2 tablespoons dry white wine
2 teaspoons lemon juice
1 teaspoon minced onion

1 teaspoon prepared horseradish
1 teaspoon prepared mustard
½ teaspoon salt
¼ teaspoon white pepper
⅛ teaspoon cayenne
6 French rolls
⅓ cup shredded Gruyere cheese

1. Pick over crab meat and remove any bits of shells or cartilage. Combine with shrimp and cream cheese and blend well. Add almonds, white wine, lemon juice, onion, horseradish, mustard, salt, pepper, and cayenne.
2. Remove top third from each roll and scoop out inside, being careful not to puncture shell. Spoon mixture evenly into 6 shells. Sprinkle cheese over top of filling. Place tops on rolls.
3. Place 2 rolls at a time on paper towels or paper plates in oven. Cook on HIGH for 1 to 1½ minutes, or until filling is piping hot and cheese has melted.

Sausage and Pepper Hero

4 servings

4 Italian sausages

½ cup prepared barbecue sauce

1 green pepper, seeded and cut in strips

4 hero rolls

1. Place a layer of paper towels in an 8-inch square dish. Place sausage on towels. Cover with paper towels.
2. Cook on HIGH for 8 minutes, turning halfway through cooking period. Drain off fat and reserve.
3. Cook barbecue sauce with pepper strips in a 2-cup measure on HIGH for 2 minutes.
4. Split hero rolls almost in half. Place 1 cooked sausage in each roll. Top with one-quarter of the sauce and peppers. Wrap each roll in waxed paper and twist ends of paper.
5. Cook, one at a time, on HIGH for 1 to 1½ minutes, or until rolls are piping hot.

Hot Dogs

1 serving

1 frankfurter roll

Prepared mustard

Pickle relish

1 frankfurter

1. Spread roll with mustard and relish. Place hot dog in roll. Wrap in waxed paper and twist ends securely.
2. Cook one hot dog on HIGH for 45 seconds. Cook 2 hot dogs at a time on HIGH for 1½ minutes; cook 3 hot dogs on HIGH for 2½ to 3 minutes; cook 4 hot dogs on HIGH for 3 to 3½ minutes, or until piping hot.

Beans and Things

6 servings

6 slices bacon, cut in pieces

1 medium onion, chopped

4 frankfurters, cut in small pieces

1 teaspoon prepared mustard

1 teaspoon catsup

1 can (1 pound) pork and beans in tomato sauce

6 slices toast

1. Put bacon pieces in 1½-quart casserole.
2. Cook, covered, on HIGH for 2 to 3 minutes. Stir.
3. Add chopped onion.
4. Cook, covered, on HIGH for 2 minutes. Stir well.
5. Add frankfurters. Cook, covered, on HIGH for about 3 minutes. Add mustard, catsup, and beans and mix well.
6. Cook, covered, on HIGH for about 4 minutes, or until hot, stirring once during cooking period. Serve over toast.

Sausage and Pepper Hero, Hot Dogs

Asparagus Egg Specials

6 servings

6 slices white bread	Salt and pepper to taste
2 tablespoons soft butter or margarine	1 can (8 ounces) tomato sauce
3 hard-cooked eggs, sliced	¼ teaspoon sugar
24 cooked green asparagus spears (see page 94)	½ teaspoon oregano
	2 tablespoons slivered almonds

1. Toast bread and spread lightly with butter.
2. Arrange egg slices on toast and top each toast slice with 4 asparagus spears. Sprinkle with salt and pepper.
3. Combine tomato sauce, sugar, and oregano.
4. Place toast slices in a 9-inch square dish. Pour tomato sauce over and sprinkle with nuts.
5. Cook on HIGH for 2 to 3 minutes, or until heated through.

Chicken Crumpets

8 servings

4 tablespoons butter or margarine	3 tablespoons sherry
¼ cup all-purpose flour	2 cups diced cooked chicken
1 teaspoon salt	Paprika
⅛ teaspoon pepper	4 crumpets or English muffins
2 cups milk	8 baked ham slices

1. Place butter in a 1-quart measure. Cook on ROAST for 30 seconds, or until butter is melted.
2. Stir in flour, salt, and pepper. Cook on HIGH for 30 seconds.
3. Stir in milk. Cook on HIGH for 3 to 4 minutes, stirring twice during cooking time.
4. Remove from oven and stir in sherry, chicken, and a dash of paprika.
5. Split, toast, and butter crumpets or English muffins. Place each on a paper plate.
6. Place 1 slice of ham on each crumpet half. Divide chicken mixture over ham slices and sprinkle lightly with paprika.
7. Cook 4 at a time on HIGH for about 2 minutes, or until heated through.

Chicken Tacos

10 to 12 servings

½ cup chopped onions	¼ teaspoon garlic salt
2 tablespoons butter or margarine	10 to 12 fully cooked taco shells
2 cups coarsely diced cooked chicken	Grated Cheddar cheese
1 can (7½ ounces) taco sauce	Shredded lettuce
¼ teaspoon salt	Chopped fresh tomatoes

1. Cook onion and butter in a 1-quart measure on SAUTE for about 3 minutes.
2. Add chicken, taco sauce, salt, and garlic salt. Cover with a paper towel.

3. Cook on HIGH for about 5 minutes, or until mixture thickens and is hot.
4. Spoon hot filling into taco shells. Serve immediately with side dishes filled with cheese, lettuce, and tomato to sprinkle over hot filling.

Turkey Glory   6 servings

6 slices cooked turkey
6 slices white bread, toasted and buttered
1 package (8 ounces) cream cheese, at room temperature

1 cup milk
½ cup grated Parmesan cheese
¼ teaspoon garlic salt
½ cup sliced stuffed olives
Paprika

1. Place turkey on toast.
2. Put softened cream cheese in a 1-quart measure. Gradually add milk to cheese, mixing until well blended.
3. Cook on BAKE for 3 minutes.
4. Stir. Cook on BAKE for 2 minutes.
5. Remove from oven and stir in Parmesan cheese, garlic salt, and olives.
6. Cover each sandwich with sauce; sprinkle with paprika.

Sloppy Joe Sandwich   6 servings

1 pound ground beef
½ cup chopped onions
½ cup chopped green pepper
½ teaspoon paprika
1 can (8 ounces) tomato sauce

1 teaspoon salt
Pinch of sugar
Freshly ground pepper to taste
Toasted hamburger buns

1. Crumble beef in a 2-quart casserole. Add onion, pepper, and paprika.
2. Cook, uncovered, on HIGH for 4 minutes, or until meat loses its red color. Stir once during cooking time.
3. Break up meat with a fork. Add remaining ingredients except buns and blend well.
4. Cook, covered, on HIGH for 10 minutes, stirring occasionally.
5. Spoon onto bottom half of toasted hamburger buns; cover with top half.

Note: The Sloppy Joe mixture can be made up well in advance and kept in the refrigerator. To serve, remove any congealed fat on top of mixture. Spoon desired amount of meat on hamburger buns or hard rolls, spreading mixture out to edges. Place single serving on a small plate. Heat for 1 to 1½ minutes, until mixture is piping hot.

111

Cheeseburgers
4 servings

1 pound ground beef
Salt and pepper

4 hamburger buns, toasted
4 slices process American cheese

1. Season ground beef to taste with salt and pepper. Shape into 4 patties. Place in an 8-inch square baking dish.
2. Cook, covered with waxed paper, on REHEAT for 2 minutes.
3. Turn patties over. Cook, covered, on REHEAT for 2 minutes, or to the desired degree of doneness.
4. Place 1 patty on each hamburger bun. Top with a slice of cheese. Place each bun on a small paper plate.
5. Cook 1 or 2 at a time on REHEAT for 1 minute, or until cheese melts.

Witches' Brew Heroes
6 servings

1 medium onion, chopped
1 small green pepper, seeded and chopped
1 clove garlic, peeled and halved
1 pound ground beef, crumbled
1 can (1 pound) whole tomatoes

½ teaspoon chili powder
¼ teaspoon salt
¼ teaspoon sugar
¼ teaspoon hot-pepper sauce
⅛ teaspoon ground cumin
6 hero rolls

1. Combine onion, green pepper, garlic, and beef in a 2-quart casserole.
2. Cook, uncovered, on HIGH for 4 minutes, or until meat loses its red color. Stir once to break up meat.
3. Break up large meat chunks with a fork. Mash tomatoes with a fork so that tomatoes are in small chunks. Add to meat mixture with chili powder, salt, sugar, hot-pepper sauce, and cumin.
4. Cook, covered, on HIGH for 10 minutes, stirring once.
5. Remove garlic. Serve on hero rolls.

Paul Bunyans
6 servings

1 pound ground beef
½ cup finely chopped onion
½ cup finely chopped green pepper
½ cup chopped pitted ripe olives
1 can (6 ounces) tomato paste
1 teaspoon salt

Freshly ground pepper to taste
½ teaspoon poultry seasoning
½ teaspoon hot-pepper sauce
½ teaspoon chili powder
1 teaspoon Worcestershire sauce
Toasted hamburger buns

1. Crumble beef into a 2-quart casserole. Add onion and green pepper.
2. Cook, uncovered, on HIGH for 4 minutes, or until meat has lost its red color. Stir once during cooking time.
3. Break up meat chunks with a fork. Add olives, tomato paste, salt, pepper, poultry seasoning, hot-pepper sauce, chili powder, and Worcestershire. Stir well. Add 1 to 2 tablespoons of water if mixture is dry.
4. Cook, covered, on HIGH for 10 minutes, stirring once.
5. Serve immediately on toasted buns.

Turkey Gobble-Up

6 servings

6 English muffins
Butter
6 slices bacon
1 large avocado
¼ cup mayonnaise
¼ cup dairy sour cream
1 tablespoon lemon juice

Dash of hot-pepper sauce
12 slices (1 ounce each) cooked turkey breast
12 slices tomato
1 jar (8 ounces) pasteurized process cheese spread

1. Split, toast, and butter English muffins.
2. Cut bacon strips into 4 pieces each. Place on a paper towel in an oblong baking dish. Cover with a paper towel. Cook on HIGH for 2 to 3 minutes.
3. Peel, seed, and mash avocado. Combine with mayonnaise, sour cream, lemon juice, and hot-pepper sauce.
4. Spread avocado mixture generously on each muffin half. Arrange 1 slice each turkey and tomato on each muffin half. Spread 1 tablespoon cheese on each tomato. Top each with 2 pieces of cooked bacon.
5. Place 6 halves in an oblong baking dish. Cook on ROAST for 1½ to 2 minutes, or until cheese is hot and bubbly.

Ham and Asparagus Cheese Sandwiches

4 servings

4 slices bread, toasted
4 slices thinly sliced ham

8 slices Swiss cheese
1 can (15 ounces) asparagus spears

1. Place toasted bread in a baking dish.
2. Cover each slice with 1 slice ham, 1 slice cheese, and 5 asparagus spears.
3. Cook on HIGH for 3 to 4 minutes, or until cheese is piping hot.
4. Garnish with paprika if desired.

Beverages

The microwave oven saves time when making or reheating beverages right in the cup. Use glass or pottery mugs, styrofoam or paper cups, and make single servings ready to drink. If the beverage cools, the microwave oven will restore the serving temperature in a flash. The temperature of the liquid before heating will make a difference in the heating time. Water from the tap or milk from the refrigerator will take longer to heat than most beverages. Beverages taste their best when heated to a temperature of between 150°F. and 160°F.

Remember that milk boils very easily and quickly, so when making beverages with milk, do not fill the container to the brim.

Beverages — How to Heat

1. Save time by making or reheating beverages right in the cup. Use glass or pottery mugs without silver or other metal trim.
2. Microwave on HIGH.
3. Most beverages taste best if heated to a temperature of between 150°F. and 160°F.
4. Watch milk carefully so that it does not boil over.
5. The temperature of the liquid before heating will make a difference in final heating time. Water from the cold tap or milk from the refrigerator will take longer to heat.

Liquid	6-ounce Cup	Time in Minutes	8-ounce Cup	Time in Minutes
Water	1	1 to 1¼	1	1½ to 2
	2	1¾ to 2	2	3 to 3¼
Milk	1	2½	1	2¾ to 3
	2	2¾ to 3	2	3¼ to 3½
Reheating coffee	1	1 to 1½	1	1¼ to 1½
	2	2 to 2¼	2	2 to 2½

Hot Tea   4 servings

4 tea bags

1. Place 3 cups water in a 4-cup measuring cup.
2. Heat on HIGH for 5 minutes.
3. Add tea bags. Remove when desired strength of tea is obtained.

Iced Tea   4 servings

6 tea bags, or 2½ to 3 tablespoons   **3 to 4 tablespoons sugar**
instant tea   **Ice cubes**

1. Place 3 cups water in a 4-cup measuring cup.
2. Heat on HIGH for 5 minutes.
3. Add tea.
4. When desired strength is obtained, take out tea bags and stir in sugar until dissolved.
5. Serve over ice in tall glasses.

Café Au Lait   4 servings

4 teaspoons instant coffee   **Sugar (optional)**
2 cups milk

1. Place coffee and 1 cup water in a 4-cup measuring cup.
2. Stir in milk.
3. Cook on HIGH for 4 to 5 minutes, or until piping hot.
4. Sweeten to taste, if desired.

116

Café Calypso
6 servings

4 cups milk
⅓ cup instant coffee
¼ cup brown sugar

½ cup heavy cream, whipped
Ground nutmeg

1. Place ⅓ cup water in a 2-quart casserole.
2. Heat on HIGH for 1 minute.
3. Add milk, coffee, and sugar.
4. Cook on HIGH for 4 to 5 minutes.
5. Serve hot, topped with whipped cream and a dash of nutmeg.

Viennese After-Dinner Coffee
8 servings

6 teaspoons instant coffee

¼ cup heavy cream, whipped

1. Combine coffee and 3 cups water in a 4-cup measuring cup.
2. Cook on HIGH for 4 to 5 minutes, or until piping hot.
3. Serve in demitasse cups.
4. Top each serving with whipped cream.

Spicy Orange Coffee
6 servings

1 tablespoon sugar
6 whole cloves
2 pieces (1½ inches each) stick
   cinnamon

Peel of small orange, in strips
1 tablespoon instant coffee

1. Combine all ingredients with 1½ cups water in a 2-cup measuring cup.
2. Heat on HIGH for 4 to 5 minutes, or until piping hot.
3. Strain into demitasse cups.

Coffee Cream Punch
8 servings

6 tablespoons instant coffee
1½ pints vanilla ice cream

Ground nutmeg

1. Combine coffee and 4 cups water in a 1½-quart casserole.
2. Cook on HIGH for 5 to 6 minutes, or until piping hot.
3. Meantime, place ice cream in a 3-quart bowl.
4. Pour hot coffee over the ice cream; stir until melted.
5. Ladle into cups and sprinkle each serving with nutmeg.

Mexican Chocolate
4 servings

½ cup semisweet chocolate bits
1 tablespoon instant coffee
½ teaspoon vanilla extract

¼ teaspoon ground cinnamon
2 cups milk

1. Place chocolate, coffee, and ½ cup water in a 4-cup measuring cup.
2. Heat on HIGH for 2 minutes.
3. Remove from oven and stir in remaining ingredients.
4. Heat on HIGH for 4 minutes.

Hot Instant Tea

4 servings

5 to 6 teaspoons instant tea

1. Mix tea with 3 cups water in a 4-cup measuring cup.
2. Heat on HIGH for 5 minutes.

Instant Coffee

4 servings

4 to 5 teaspoons instant coffee

1. Combine coffee and 3 cups water in a 1-quart measuring cup.
2. Cook on HIGH 4 to 5 minutes, or until hot. To develop a richer flavor, let stand for 2 minutes.

Instant Cocoa

1 serving

¾ cup milk **Marshmallows**
2 teaspoons instant cocoa mix

1. Combine milk and cocoa in a 2-cup measuring cup.
2. Cook on ROAST for 2 minutes.
3. Add marshmallows and serve.

Mulled Cider

4 servings

3 cups apple cider **½ teaspoon whole allspice**
3 tablespoons brown sugar **½ teaspoon whole cloves**
⅛ teaspoon salt **1 stick cinnamon**
Dash of ground nutmeg **2 orange slices, cut in half**

1. Place apple cider, brown sugar, salt, and nutmeg in 1-quart measuring cup.
2. Tie allspice, cloves, and cinnamon in cheesecloth and drop into cider.
3. Cook on HIGH for 4 to 5 minutes. Let stand 5 minutes.
4. Remove spice bag and serve hot, garnished with orange slices.

Hot Spiced Cranberry Punch

6 servings

1½ cups cranberry juice cocktail **1 can (6 ounces) frozen lemonade**
4 whole cloves **concentrate, thawed**
1 piece (2 inches) stick cinnamon **3 orange slices, cut in half**
3 tablespoons sugar **6 maraschino cherries**

1. Combine cranberry juice, cloves, and cinnamon stick with 1½ cups water in 1-quart measuring cup.
2. Cook on HIGH for 4 minutes.
3. Cover and let stand 1 minute.
4. Remove spices.
5. Stir in sugar until dissolved.
6. Blend in lemonade.
7. Cook on HIGH for 3 minutes.
8. Serve hot, garnished with half an orange slice and a maraschino cherry on a toothpick.

Mulled Wine

8 to 10 servings

1 cup sugar

2 pieces (1 inch each) stick cinnamon

1 lemon, sliced

24 whole cloves

4 cups orange juice

1 quart Burgundy wine

1. Combine sugar, cinnamon, lemon, and cloves with ½ cup water in a 3-quart casserole.
2. Heat on HIGH for 2 minutes.
3. Add orange juice and Burgundy.
4. Heat on HIGH for 7 to 8 minutes.
5. Garnish with lemon or pineapple slices, if desired.

Hot Toddy

1 serving

1 teaspoon sugar

1 piece (1 inch) stick cinnamon

1 slice lemon, studded with 2 cloves

2 ounces bourbon

1. Combine sugar, cinnamon, and lemon slice with ½ cup water in a 1-cup measuring cup.
2. Cook on HIGH for 2½ minutes.
3. Meantime, place bourbon in a serving cup or mug.
4. Remove hot mixture from oven and pour over bourbon. Stir and serve.

Mulled Pineapple Juice

10 servings

1 can (46 ounces) pineapple juice

1 piece (2 inches) stick cinnamon

⅛ teaspoon ground nutmeg

⅛ teaspoon ground allspice

Dash ground cloves

1. Combine all ingredients in a 2-quart casserole.
2. Heat on HIGH for 7 to 8 minutes.

Spicy Apple Nog

5 to 6 servings

2 eggs, separated

¼ cup sugar

½ teaspoon salt

½ teaspoon ground cinnamon

Dash of ground nutmeg

⅔ cup apple juice

3 cups milk

½ cup heavy cream, whipped

1. Place egg yolks in a 2-quart casserole. Beat lightly with a fork.
2. Stir in sugar, salt, cinnamon, nutmeg, and apple juice until well blended. Stir in milk.
3. Cook on ROAST for 7 to 8 minutes, or until piping hot.
4. Meantime, beat egg whites in a 2-quart mixing bowl.
5. Remove milk mixture from oven and pour quickly over egg whites, stirring rapidly.
6. Top each serving with a mound of whipped cream.

Desserts:
Fruits, Puddings, Pies

Using the microwave oven, fruits and custards—or both in combination—can be prepared easily and quickly. Poached fruits especially, by retaining their color and shape, make an attractive serving. Custard desserts should be removed from the oven when the center of the custard is nearly set; it will continue cooking after being removed. For cooking fresh fruits, the HIGH setting is used; canned fruits need only be heated, since they have been cooked at the time of canning. Most such desserts are cooked uncovered. Dried fruits, however, cook best when covered, as they need to retain moisture in order to plump.

Pie crusts cook beautifully on ROAST. They should be pricked before baking and cooled before filling. Two-crust pies are cooked first in the microwave oven, then transferred to a conventional oven for browning. Even leftover slices of pie can be quickly reheated.

Cherry Cheese Pie

1 9-inch pie

⅓ cup butter or margarine
1¼ cups graham cracker crumbs
¼ cup all-purpose flour
Sugar
  1 package (8 ounces) cream cheese, softened

1 egg, lightly beaten
1 cup dairy sour cream, divided
1¾ teaspoons vanilla extract, divided
1 can (21 ounces) prepared cherry pie filling

1. Put butter in a 9-inch pie plate. Melt on ROAST for 2 to 3 minutes.
2. Add graham cracker crumbs, flour, and 4 teaspoons sugar to melted butter in pie plate and blend well. Press mixture evenly over bottom and up sides of pie plate. Set aside.
3. Beat together softened cream cheese and ⅓ cup sugar until well blended. Add egg, ¼ cup sour cream, and ¾ teaspoon vanilla. Beat until light and fluffy. Pour into prepared graham cracker crust.
4. Cook on REHEAT for 4 minutes.
5. Remove from oven and cool 8 minutes on a cooling rack.
6. Beat together ¾ cup sour cream with 2 tablespoons sugar and 1 teaspoon vanilla. Spoon carefully over top of cooked cheese mixture.
7. Cook on REHEAT for 4 minutes, or just until sour cream is set.
8. Spoon cherry pie filling around edge of pie.
9. Chill thoroughly in refrigerator before serving.

Apple Pie

1 9-inch pie

7 medium apples
¾ cup sugar
2 tablespoons all-purpose flour
⅛ teaspoon salt
1 teaspoon ground cinnamon

¼ teaspoon ground nutmeg
1 recipe pie crust
1 to 2 teaspoons lemon juice
2 tablespoons butter

1. Pare and slice apples. Mix in a bowl with sugar, flour, salt, cinnamon, and nutmeg. Set aside.
2. Roll out half of the pie crust and fit in the bottom of a 9-inch pie dish. Put apples in pie crust. Sprinkle lemon juice over top if apples are not too tart. Dot with butter. Roll out remainder of pie crust and fit over apples. Seal edges and cut slits in top of pie.
3. Cook on HIGH for 10 minutes, or until apples are tender.
4. While apples are cooking, preheat conventional oven to 450° F.
5. When apples are tender, bake pie in conventional oven 12 to 14 minutes, or until crust is golden brown.
6. Serve warm or cold.

Note: The kind of pie crust that is used makes a difference in the browning time. Packaged pie crust mix browns faster than homemade.

Cherry Cheese Pie

Pie Crusts — How to Bake

1. Prepare pie crust recipe or box mix as directed on package.
2. Flute edge; prick bottom and sides of crust with fork.

Pie Crust	Minutes to Cook	Setting	Special Notes
Unbaked pie crust	7 to 8	ROAST	
Box mix pie crust	7 to 8	ROAST	
Frozen, prepared pie crust	6 to 6½	HIGH	Remove from metal container and place in glass pie plate.

Pies — How to Bake

1. Place pie in glass pie plate.
2. Microwave according to chart or until filling is warmed through.

Pie	Minutes to Cook	Setting	Special Notes
Fruit			
Fresh, 2-crust, 9-inch pie	7 to 8	HIGH	After microwave cooking is completed, transfer to preheated conventional oven for 10 to 15 minutes at 450° F.
Frozen, 9-inch pie	15	HIGH	After microwave cooking is completed, transfer to preheated conventional oven for 10 to 12 minutes at 425° F.
Custard			
1-crust, 9-inch pie	4 to 4½	ROAST	

Pecan Pie   1 9-inch pie

¼ cup butter or margarine
⅓ cup brown sugar
1 cup corn syrup
3 eggs, lightly beaten

1 teaspoon vanilla
Pinch of salt
1 cup pecan halves
1 baked 9-inch pastry shell

1. Place butter in a medium-size glass bowl. Cook on HIGH for 1½ minutes, or until butter is melted.
2. Stir in remaining ingredients except pastry shell. Pour mixture into shell.
3. Cook on BAKE for 10 to 12 minutes, or until custard center is set.
4. Cool and serve with whipped cream, if desired.

Pumpkin Pie

1 9-inch pie

2 eggs, lightly beaten
1½ cups solid-pack cooked pumpkin
¾ cup sugar
½ teaspoon salt
1 teaspoon ground cinnamon

½ teaspoon ground ginger
¼ teaspoon ground cloves
1 can (14½ ounces) evaporated milk
1 9-inch baked pastry shell

1. Combine eggs, pumpkin, sugar, salt, and spices and blend well. Stir in milk and make a smooth mixture.
2. Remove ⅔ cup of this mixture and set aside. Pour remaining mixture into baked pastry shell.
3. Cook on SIMMER for 15 minutes. Turn pie and cook on SIMMER for 20 minutes.
4. Let pie stand 20 minutes to continue cooking.
5. Cool before cutting. Serve with flavored whipped cream, since top of pie may have a rough appearance.

Note: Pour reserved pumpkin mixture into custard cups, filling them three-quarters full. Cook on SIMMER for 5 to 6 minutes, or until a knife inserted near the center comes out clean.

Stewed Apricots

6 servings

½ pound dried apricots
1 cup white raisins
Juice of 1 lemon, or 2 tablespoons
   lemon juice

½ cup sugar
1 can (11 ounces) mandarin oranges,
   drained

1. Rinse apricots and raisins in water. Drain.
2. Put apricots and raisins in a 1½-quart casserole. Add 1½ cups water, and cook, uncovered, on HIGH for 5 minutes.
3. Add lemon juice, sugar, and mandarin oranges. Cook on HIGH for 5 minutes. Let stand 2 or 3 minutes before serving.

Baked Maple Bananas

4 servings

2 tablespoons butter or margarine
3 tablespoons maple syrup

4 bananas
Lemon juice

1. Place butter in a medium-size baking dish. Cook on HIGH for 1 minute, or until butter is melted.
2. Add maple syrup and mix.
3. Place peeled bananas in dish and spoon butter mixture over, so that bananas are well coated. Cook on HIGH for 1 minute. Turn bananas. Cook on HIGH for 1½ minutes.
4. Remove from oven and sprinkle with lemon juice.

Honeyed Blueberries

4 to 6 servings

3 cups bran flakes	1 teaspoon ground cinnamon
½ cup honey	½ teaspoon ground nutmeg
¼ cup sugar	2 cups fresh blueberries

1. In a bowl, combine bran flakes, honey, sugar, cinnamon, and nutmeg.
2. Grease an 8-inch square baking dish. Spread half the bran flakes mixture on the bottom. Cover with half the blueberries. Cover blueberries with remaining bran flakes and top with remaining blueberries.
3. Cook, covered, on HIGH for 4 minutes. Serve hot.

Baked Grapefruit

4 servings

2 grapefruit	4 teaspoons honey
4 teaspoons dry sherry	

1. Cut grapefruits in half. Loosen each section of grapefruit. Top each half with 1 teaspoon dry sherry and 1 teaspoon honey.
2. Cook on HIGH for 5 minutes, or until grapefruits are very hot.

Date-Filled Pears

4 servings

4 fresh Bartlett pears	3 tablespoons butter or margarine
½ cup pitted dates, cut in small pieces	⅓ cup dry vermouth
2 tablespoons light brown sugar	

1. Cut pears in half. Peel and core. Place in baking dish, cut-side up.
2. Fill pear centers with cut dates. Sprinkle with brown sugar and dot with butter.
3. Pour vermouth over pears and cook, uncovered, on HIGH for 8 minutes, basting once or twice during cooking time. Let stand 2 or 3 minutes.

Baked Custard

5 servings

3 eggs	1¾ cups milk
¼ cup sugar	1 teaspoon vanilla extract
¼ teaspoon salt	Ground nutmeg

1. Beat eggs until fluffy. Add sugar and salt and continue beating until thick and lemon-colored. Beat in milk and vanilla.
2. Divided mixture among five 6-ounce custard cups. Sprinkle with nutmeg.
3. Arrange cups in a glass baking dish. Fill dish about half full with boiling water. Cook on SIMMER for 9 to 10 minutes.
4. Remove and let stand 5 minutes.

Note: This custard can be baked in an 8-inch cake dish. The consistency will be a little softer. If desired, bake on BAKE for 25 to 30 minutes, or until almost set.

Quick Crème Brûlée

4 servings

1 package (3¼ ounces) pudding mix, vanilla flavor	Brown sugar

1. Prepare pudding mix according to preceding directions. Pour cooked pudding into a flat baking dish. Cool well and chill.
2. Sift a layer of brown sugar, about ⅛ inch thick, over top of pudding, making sure to cover entire top.
3. Place under a hot broiler and broil until sugar melts and bubbles. Watch carefully so that sugar does not burn.

Pudding and Pie Filling Mix — How to Cook

1. Prepare mix as directed on package. Blend well.
2. Cook mix in a 1-quart glass measure or mixing bowl, stirring once during cooking time.
3. Pour into serving dishes.
4. Chill before serving.

	Size of Package	Minutes to Cook	Setting
Pudding and pie filling mix	3¼-ounce, 4 servings	5 to 7	HIGH
	5½-ounce, 6 servings	8 to 9	HIGH
Golden egg custard	3 ounces, 4 servings	8 to 10	ROAST
Tapioca	3¼ ounces, 4 servings	5 to 7	ROAST

Old-Fashioned Indian Pudding   4 to 6 servings

2 cups milk, divided
¼ cup yellow cornmeal
2 tablespoons sugar
½ teaspoon salt
½ teaspoon ground cinnamon
¼ teaspoon ground ginger

1 egg, beaten
¼ cup molasses
1 tablespoon melted butter or margarine
Vanilla ice cream

1. Pour 1½ cups milk into a 1½-quart casserole. Heat on SIMMER for 5 minutes.
2. Combine cornmeal, sugar, salt, cinnamon, and ginger. Stir into hot milk.
3. Cook, uncovered, on SIMMER for 4 minutes. Stir well.
4. Beat together egg, molasses, and butter. Stir a small amount of hot milk mixture into egg mixture. Return to casserole. Stir well.
5. Cook, uncovered, on SIMMER for 6 minutes.
6. Pour remaining ½ cup cold milk carefully over top of pudding. Do not stir. Cook, uncovered, on SIMMER for 3 minutes, or until set.
7. Let stand 10 to 15 minutes before serving.
8. Serve warm, topped with vanilla ice cream.

Party-Pretty Pudding

8 servings

2 packages (3¼ ounces each) pudding
  mix, vanilla flavored

2 pints fresh strawberries, hulled, sliced,
  and sweetened

1. Prepare pudding according to Pudding and Pie Filling Mix Chart. Cool well.
2. In a glass serving dish, spoon one-third of the cooled pudding. Cover with half the strawberries.
3. Repeat with a second layer of pudding and the remaining strawberries. Spoon the remaining third of the pudding on top of the strawberries. Top with sweetened whipped cream, if desired.

Note: If desired, the layers may be made in individual serving dishes.

Baked Apples Supreme

6 servings

6 baking apples
Lemon juice
½ cup slivered almonds
¼ cup raisins

¼ cup brown sugar
2 teaspoons ground cinnamon
6 teaspoons butter or margarine

1. Wash apples and remove core, making a generous cavity in each apple. Remove thin circle of peel around cavity and sprinkle with lemon juice.
2. Mix together almonds, raisins, brown sugar, and cinnamon. Fill cavities with mixture and place each apple in a small custard dish. Put 2 tablespoons of water in dish around apple. Dot each apple with 1 teaspoon butter.
3. Cook on HIGH for 6 minutes.
7. Let stand 2 to 3 minutes before serving.

Quick Peach Delight

4 servings

4 large canned peach halves
1¼ teaspoons butter or margarine

4 teaspoons brown sugar
Vanilla ice cream

1. Drain peaches thoroughly. Place in a 1-quart baking dish. Put ¼ teaspoon butter in center of each peach. Sprinkle 1 teaspoon brown sugar on each peach half.
2. Bake, uncovered, on HIGH for 3 minutes, or until piping hot.
3. Serve warm with a small scoop of ice cream in center of each peach half.

Quick Applescotch

4 servings

1 can (1 pound) pie-sliced apples
½ package (6 ounces) butterscotch
flavored morsels
1 tablespoon quick-cooking tapioca
½ tablespoon lemon juice
¼ cup all-purpose flour
¼ cup sugar
½ teaspoon ground cinnamon
¼ cup firm butter or margarine

1. Combine apples, butterscotch morsels, and tapioca in a 1-quart casserole. Sprinkle lemon juice over the top.
2. Combine flour, sugar, and cinnamon in a small bowl. Cut in butter with a pastry blender or two knives until mixture resembles cornmeal. Sprinkle over top of apple mixture.
3. Cook, uncovered, on BAKE for 12 minutes, or until piping hot.
4. Serve warm with heavy cream or ice cream, if desired.

Cranberry Apple Crunch

6 servings

1 cup sugar
2 cups chopped cranberries
2 cups chopped apples
1 cup quick-cooking rolled oats
½ cup firmly packed brown sugar
⅓ cup all-purpose flour
½ teaspoon salt
¼ cup butter or margarine
½ cup chopped nuts
Whipped cream

1. Combine sugar, 1 cup water, cranberries, and apples in a buttered 2-quart casserole or baking dish.
2. Cook, covered, on HIGH for 10 minutes.
3. Mix together oats, sugar, flour, and salt. Cut in butter with two knives to make a coarse mixture. Stir in nuts. Sprinkle over top of cranberry mixture.
4. Cook, covered, on HIGH for 5 minutes.
5. Cook, uncovered, on HIGH for 4 minutes, or until apples are done.
6. Let stand 3 to 4 minutes before serving. Serve with whipped cream.

Apple Betty

6 servings

⅓ cup melted butter or margarine
2 cups fresh bread crumbs
6 cups sliced, peeled, and cored
cooking apples
½ cup firmly packed brown sugar
½ teaspoon ground nutmeg
¼ teaspoon ground cinnamon
1 tablespoon grated lemon peel
(optional)
2 tablespoons lemon juice

1. Toss melted butter with bread crumbs. Put one-third of the buttered bread crumbs in a 2-quart casserole.
2. Combine apples with brown sugar, nutmeg, cinnamon, and lemon peel. Put half of the apple mixture on bread crumbs layer. Cover with one-third of the crumbs. Add remaining apples.
3. Combine lemon juice and ¼ cup water. Pour over apples. Top with remaining buttered crumbs.
4. Cook, covered, on REHEAT for 9 minutes.
5. Remove cover and cook on REHEAT for 15 minutes, or until apples are tender.

Fresh Rhubarb Betty

6 to 8 servings

6 cups diced fresh rhubarb
1¼ cups sugar
2½ tablespoons quick-cooking tapioca
1 teaspoon grated lemon peel

1 tablespoon grated orange peel
2¾ cups soft bread cubes
⅓ cup butter or margarine
1 teaspoon vanilla extract

1. Combine rhubarb, sugar, tapioca, lemon peel, and orange peel in a bowl. Set aside.
2. Put bread cubes in a bowl.
3. Place butter in a 1-cup measure. Cook, covered, on HIGH for 45 seconds, or until butter melts.
4. Pour butter over bread cubes; add vanilla and toss lightly.
5. In a 1½-quart casserole make alternate layers of rhubarb and bread-cube mixture, ending with buttered bread cubes.
6. Cook, covered, on HIGH for 5 minutes, or until rhubarb is cooked.
7. Serve warm or chilled.

Butterscotch Sauce

1 cup

½ cup sugar
½ cup firmly packed dark brown sugar
½ cup light cream

1 teaspoon vanilla extract
2 tablespoons butter
⅛ teaspoon salt

1. Combine all ingredients in a 2-cup measuring cup.
2. Cook, uncovered, on ROAST for about 4 minutes, or until sauce is well blended and hot, stirring once.
3. Serve warm over ice cream.

Fancy Chocolate Sauce

2 cups

1 package (12 ounces) semisweet chocolate bits
2 squares (2 ounces) unsweetened chocolate

1 cup heavy cream
3 tablespoons brandy

1. Combine chocolate bits and unsweetened chocolate in a small mixing bowl.
2. Cook, covered, on BAKE for 5 minutes, or just until chocolate melts. Watch carefully during last minute of cooking so that it does not burn.
3. Stir in cream with a wire whisk to make a smooth paste.
4. Cook, covered, on BAKE for 1 to 1½ minutes, or until piping hot.
5. Stir in brandy.
6. Serve hot over vanilla ice cream or cake squares.

Brandied Strawberry Sauce   1½ cups

1 pint fresh strawberries   2 tablespoons lemon juice
1 cup sugar   2 tablespoons brandy
1 tablespoon cornstarch

1. Clean and crush strawberries.
2. Combine sugar and cornstarch in a 1-quart mixing bowl. Stir in lemon juice and crushed strawberries.
3. Cook, covered, on REHEAT about 5 minutes, or until mixture comes to a boil and is clear.
4. Cool slightly. Stir in brandy. Chill well before serving.

Vanilla Mousse with Strawberry Sauce   8 to 10 servings

2 envelopes unflavored gelatin   1 pint heavy cream, whipped
1½ cups sugar, divided   1 pint fresh strawberries
1½ cups milk   2 tablespoons cornstarch
2 eggs, separated   ½ cup lemon juice
1 tablespoon vanilla extract   2 tablespoons butter

1. Combine gelatin and 1 cup sugar in a mixing bowl and blend well. Stir in milk.
2. Cook, uncovered, on HIGH for 5 minutes, or until hot.
3. Beat egg yolks in a small dish. Gradually stir in a small amount of hot milk mixture. Add to large bowl of hot milk mixture. Blend well.
4. Cook, uncovered, on BAKE for about 4 minutes, or just until bubbles form around edge of bowl. Do not overcook or mixture will curdle.
5. Stir in vanilla. Place bowl in a pan or bowl of ice water. Cool until custard mounds when dropped from a spoon.
6. Beat egg whites until stiff but not dry. Fold into custard mixture. Fold in whipped cream. Turn mixture into a 2-quart mold. Chill in the refrigerator until set.
7. To make the sauce: Clean and hull berries. Reserve 1 cup of the best berries for garnish. Force the remainder through a food mill, or blend in an electric blender. Put through a strainer to remove seeds.
8. Combine the remaining ½ cup sugar with cornstarch in a 1-quart mixing bowl. Gradually stir in 1 cup water.
9. Cook, uncovered, on HIGH for 3 to 4 minutes, or until mixture comes to a boil and is clear.
10. Stir in lemon juice, butter, and strawberry puree. Chill sauce.
11. Unmold mousse on a serving platter. Garnish with whole strawberries. Serve with chilled strawberry sauce.

Desserts:
Cakes, Cookies, Candies

Cake-baking in a microwave oven is much faster than in a conventional oven, producing a delectable cake in look, texture—and taste. Cakes should be allowed to rise slowly and evenly on BAKE, then set on HIGH. Round glass baking dishes give best results, and layers should be baked one at a time. The bottom of the dish should be lined with waxed paper if the cake is to be unmolded; grease or flour should not be used. Dishes should never be filled more than half full, as cakes rise higher than usual in a microwave oven. When the cake begins to break away from the sides of the dish, it is done.

Cookies—either wafers or bar cookies—are delightful when microwave cooked. They are cooked and cooled on the same piece of waxed paper. Their "just-baked" flavor can be later restored in seconds.

Candies are especially easy to make in a microwave oven, as syrups can be heated to a very high temperature without the danger of burning.

Cake—How to Bake

1. Prepare batter according to directions in recipe or on cake mix package.
2. Use glass dish recommended in chart.
3. If cake is to be turned out of dish, line bottom with waxed paper.
4. Fill cake pan with 2 cups prepared batter. Use remaining batter for cupcakes.
5. Bake layers one at a time.
6. Cake is done when cake begins to break away from the sides of the baking dish.
7. Cake may appear moist in spots on top of cake. As it stands it will dry.
8. Let cake stand 5 minutes after removing from oven.
9. Cake will not be browned, so turn layers out of pans and frost.

Cake or Bread	Container	First Setting and Minutes to Cook	Second Setting and Minutes to Cook
Cake mix, 17 to 18½-ounce package	8- or 9-inch round	BAKE, 3	HIGH, 3 to 4
Snackin' cake mix	8- or 9-inch round	BAKE, 6	HIGH, 3
Pound cake, 14-ounce package	9- by 5-inch loaf pan	BAKE, 8	HIGH, 3
Pineapple upside-down cake mix	9-inch round	BAKE, 7	HIGH, 3 to 4
Coffeecake mix	9-inch round	BAKE, 4	HIGH, 3 to 4
Cupcakes: 1 2 4 6	Paper cupcake liners in custard cups or microwave muffin tray. Fill ½ full.	BAKE, 45 seconds BAKE, 1½ to 2 BAKE, 2½ to 3 BAKE, 4 to 4½	
Blueberry muffin mix: 2 muffins 4 muffins 6 muffins	Paper cupcake liners in custard cups or microwave muffin tray. Fill ½ full.	BAKE, 2 BAKE, 4 BAKE, 5-5½	
Corn muffin mix: 2 muffins 4 muffins 6 muffins	Paper cupcake liners in custard cups or microwave muffin tray. Fill ½ full.	BAKE, 1½ BAKE, 3 BAKE, 4½	
Gingerbread mix	8-inch round	BAKE, 7	HIGH, 1 to 2
Nut bread mix	9- by 5-inch loaf pan	SIMMER, 9	HIGH, 3 to 4

Snow-White Frosting

1 cup sugar
½ cup water
¼ teaspoon cream of tartar

Dash of salt
2 egg whites
1 teaspoon vanilla extract

1. Combine sugar, water, cream of tartar, and salt in a 2-cup glass measure. Cook on ROAST for 4 to 5 minutes, or until mixture boils.
2. Beat egg whites in small bowl with electric mixer until soft peaks form. Gradually pour hot syrup over egg whites, beating all the time. Continue beating for about 5 minutes, or until frosting is thick and fluffy. Beat in vanilla.

Note: Makes enough frosting for two 9-inch layers.

Pumpkin Raisin-Nut Cake   12 servings

½ cup shortening	1 teaspoon salt
1 cup sugar	2½ teaspoons ground cinnamon
2 eggs, lightly beaten	½ teaspoon ground nutmeg
1 cup solid-pack cooked pumpkin	¼ teaspoon ground ginger
2 cups sifted all-purpose flour	1 cup seedless raisins
4 teaspoons baking powder	1 cup chopped nuts
1 teaspoon baking soda	

1. Cream shortening and sugar together until light and fluffy. Beat in eggs and pumpkin; beat well.
2. Sift together flour, baking powder, baking soda, salt, and spices. Stir into pumpkin mixture and blend well. Stir in raisins and nuts. Pour batter into a lightly greased 12- by 8- by 2-inch baking dish.
3. Cook on SIMMER for 9 minutes.
4. Cook on HIGH for 6 minutes, or until a toothpick inserted in center of cake comes out clean.
5. Let cool before serving.

Sour Cream Coffee Cake   1 8-inch cake

¼ cup butter or margarine	½ cup dairy sour cream
½ cup sugar	⅓ cup brown sugar, firmly packed
2 eggs	2 tablespoons flour
½ teaspoon vanilla extract	½ cup chopped nuts
1½ cups sifted all-purpose flour	⅛ teaspoon ground cinnamon
½ teaspoon baking soda	⅛ teaspoon salt
½ teaspoon baking powder	2 tablespoons butter or margarine

1. Cream together ¼ cup butter and sugar until light and fluffy. Add eggs and vanilla and beat thoroughly. Sift together flour, soda, and baking powder. Add to creamed mixture alternately with sour cream, blending well after each addition.
2. Combine remaining ingredients and mix until crumbly.
3. Spread half of the batter in an 8-inch round cake dish. Sprinkle with half of topping mix. Spread on remaining batter and sprinkle with remaining topping.
4. Cook on BAKE for 4 minutes.
5. Cook on HIGH for 3 minutes. Serve coffee cake warm.

Devil's Food Cake

2 8-inch layers

2 cups sifted all-purpose flour	½ cup cocoa
1¼ teaspoons baking soda	1 teaspoon vanilla extract
¼ teaspoon salt	½ cup buttermilk
½ cup shortening	2 eggs, lightly beaten
2 cups sugar	

1. Grease the bottoms of two 8-inch round cake dishes. Line the bottoms with 2 layers of waxed paper.
2. Sift together flour, baking soda, and salt. Set aside.
3. Cream together shortening, sugar, cocoa, and vanilla until light and fluffy.
4. Measure 1 cup water in a 2-cup measuring cup. Cook for about 2½ minutes, or until water comes to a boil. Let stand.
5. Stir boiling water, buttermilk, and eggs into creamed mixture and beat well. Add sifted dry ingredients all at once and beat well.
6. Divide mixture between prepared cake dishes.
7. Cook, uncovered, one layer at a time on BAKE for 5 minutes.
8. Cook on HIGH for 4 minutes. Remove from oven and let stand until cake is cool.
9. Turn out of dishes and cool thoroughly. Frost as desired.

Pineapple Upside-Down Cake

6 servings

2 tablespoons butter or margarine	6 to 10 maraschino cherries, well
½ cup firmly packed dark brown sugar	drained
1 can (8¼ ounces) sliced pineapple,	1 package (9 ounces) yellow cake mix
well drained, juice reserved	Whipped cream

1. Put butter and brown sugar in an 8-inch round cake dish.
2. Cook, uncovered, on SIMMER for 2 minutes, or until butter and sugar are blended.
3. Smooth mixture over bottom of pan. Arrange pineapple slices on brown sugar and dot with maraschino cherries.
4. Prepare package mix according to directions, using ⅓ cup of the pineapple juice for part of the liquid and reducing the total liquid by 1 tablespoon.
5. Pour batter carefully into pan without disturbing pineapple or sugar mixture.
6. Cook, uncovered, on BAKE for 7 minutes.
7. Cook on HIGH for 3 to 4 minutes. Remove from oven and let stand 3 minutes.
8. Invert pan on serving plate and remove pan, leaving syrup and fruit on top of cake.
9. Serve with whipped cream.

Applesauce Cake

12 servings

1 cup applesauce
⅞ cup brown sugar, firmly packed
½ cup melted butter or margarine
1¾ cups sifted all-purpose flour
1 teaspoon baking soda
½ teaspoon salt

1 teaspoon ground cinnamon
½ teaspoon ground cloves
1 teaspoon ground ginger
½ cup seedless raisins
½ cup chopped nuts

1. In a small bowl, combine applesauce, sugar, and butter. Set aside.
2. Sift flour, baking soda, salt, and spices into a large mixing bowl. Add the applesauce mixture and blend well. Stir in raisins and nuts. Pour batter into a lightly greased 12- by 8- by 2-inch baking dish.
3. Cook on BAKE for 9 minutes.
4. Cook on HIGH for 4 to 5 minutes, or until done. Cake is done when a toothpick inserted in center comes out clean.
5. Let cool before serving.

Spice Cake

12 servings

2 eggs
1 cup sugar
2 tablespoons molasses
2 cups sifted all-purpose flour
1 teaspoon ground cinnamon
1 teaspoon ground cloves

½ teaspoon ground allspice
½ teaspoon salt
2 teaspoons baking powder
1 teaspoon baking soda
1 cup buttermilk
⅔ cup cooking oil

1. Beat eggs until thick and lemon-colored. Beat in sugar and molasses until well blended.
2. Sift together flour, spices, salt, baking powder, and baking soda. Add to egg mixture alternately with buttermilk, mixing well after each addition. Stir in oil.
3. Pour batter into a lightly greased 12- by 8- by 2-inch baking dish.
4. Cook on BAKE for 9 minutes.
5. Cook on HIGH for 6 minutes, or until a toothpick inserted in center of cake comes out clean.
6. Let cool before serving.

Sticky Buns *(Illustrated on page 73)*

6 servings

⅓ cup dark brown sugar, firmly packed
3 tablespoons butter or margarine

⅓ cup chopped nuts
1 can (8 ounces) refrigerated biscuits

1. Combine brown sugar, butter, and 1 tablespoon water in an 8-inch round baking dish.
2. Cook, uncovered, on ROAST for 2 minutes, or until butter melts.
3. Stir mixture and spread over bottom of pan. Sprinkle nuts over top. Place biscuits on top of mixture.
4. Bake, uncovered, on ROAST for 4 to 5 minutes, or until biscuits are firm and no longer doughy.
5. Let stand about 2 minutes. Invert on a flat serving plate.

Raisin Applesauce Squares

16 squares

½ cup butter or margarine
⅓ cup sugar
⅓ cup brown sugar, firmly packed
¼ cup applesauce
1 cup sifted all-purpose flour
1 teaspoon baking powder
¼ teaspoon cinnamon
1 egg
1 teaspoon vanilla extract
¼ cup chopped nuts
¼ cup raisins

1. Place butter in a small glass mixing bowl. Cook on ROAST for 1 minute.
2. Stir in sugars and applesauce. Sift together flour, baking powder, and cinnamon. Stir into butter mixture. Beat in egg. Stir in vanilla, nuts, and raisins. Spread mixture in an 8-inch square glass baking dish.
3. Cook, uncovered, on HIGH for 4 minutes. Cook on HIGH for 1 to 1½ minutes, or until set.
4. Cool thoroughly. Cut into squares.

Chocolate Smoothies

16 squares

¼ cup butter or margarine, softened
¼ cup vegetable shortening
½ cup firmly packed brown sugar
1 cup all-purpose flour
½ package (6 ounces) semisweet chocolate pieces
½ cup chopped pecans

1. Combine butter, shortening, brown sugar, and flour. Press lightly on bottom of an 8-inch square glass baking dish. Bake on ROAST for 3 to 4 minutes, or until firm.
2. Remove from oven and immediately sprinkle chocolate pieces over top. Let stand 2 minutes to soften, then spread chocolate over top of crust. Sprinkle with nuts.
3. Mark into 16 squares while warm. Cut when cool; remove from dish and let cool thoroughly before serving.

Fudge

about 3 dozen pieces

1½ cups sugar
1 tablespoon butter or margarine
½ cup evaporated milk or light cream
16 marshmallows, cut in small pieces
1 package (12 ounces) semisweet chocolate pieces
1 cup chopped nuts
1 teaspoon vanilla extract

1. Combine sugar, butter, and milk in a 2-quart glass mixing bowl. Cook on ROAST for 2 to 3 minutes, or until mixture begins to boil. Remove from oven and stir well.
2. Cook on ROAST for 2 to 3 minutes, or until mixture boils and sugar is completely dissolved. Stir in marshmallows and chocolate pieces and beat until smooth. Stir in nuts and vanilla. Spread in a buttered 8-inch square baking dish.
3. Cool. Cut into 1-inch squares.

Divinity

about 3 dozen pieces

2 cups sugar
½ cup light corn syrup
⅓ cup water

2 egg whites
1 teaspoon vanilla extract
½ cup chopped nuts

1. Combine sugar, corn syrup, and water in a 3-quart glass mixing bowl. Cook on HIGH for 5 minutes, or until mixture is clear. Stir thoroughly.
2. Cook on ROAST for 5 to 7 minutes, or until a small amount of the syrup dropped in cold water forms a hard ball.
3. While syrup is cooking, beat egg whites in a large mixing bowl until stiff peaks form.
4. When syrup is ready, pour in a thin, slow stream into egg whites, beating constantly with the electric mixer. Add vanilla. Beat about 6 to 8 minutes, or until mixture is stiff and loses it's shine. Fold in nuts.
5. Drop mixture by teaspoonfuls onto waxed paper or spread in a buttered 10- by 8-inch pan. Cut into squares when cold.

Peanut Brittle

about 1 pound

2 tablespoons butter or
   margarine
½ cup water
2 tablespoons molasses

1½ cups sugar
1¼ cups shelled roasted
   peanuts, coarsely chopped
1 teaspoon baking soda

1. Combine butter, water, molasses, and sugar in a large glass mixing bowl. Heat on HIGH for 11 to 14 minutes, or until a small amount of the syrup dropped into cold water forms hard, brittle threads.
2. Butter 13- by 9-inch glass baking dish.
3. Quickly stir peanuts and baking soda into cooked mixture. Stir until soda foams and is well combined.
4. Pour mixture into baking dish. Using two forks or spoons, spread candy into a thin layer.
5. When cold, break into pieces.

Almond Brittle

about 1½ pounds

2 cups sugar
1 cup light corn syrup
½ cup water
2 cups chopped almonds

¼ teaspoon salt
1 teaspoon butter or margarine
1 teaspoon baking soda

1. Combine sugar, corn syrup, and water in a 2-quart glass mixing bowl.
2. Cook on HIGH for 16 to 18 minutes, or until a small amount of the syrup dropped into cold water forms a soft ball.
3. Stir in almonds and salt. Cook on ROAST for 12 to 15 minutes, or until a small amount of the syrup dropped into cold water forms hard, brittle threads.
4. Stir in butter and baking soda and blend well.
5. Divide mixture between two well-buttered baking sheets. Using two forks or spoons, spread candy into thin layers.
6. When cold, break into pieces.

Microwave Made-Easy Menus

The pleasure of cooking with a microwave oven is that you are free to enjoy your company—or your family—without the bother of running back and forth to the kitchen to check on food. Microwave cooking means, too, that much of the food can be prepared in advance. A microwave oven offers the added convenience of being able to stop at a halfway point with many recipes, then finishing the cooking just before dinner, or while part of the meal is being served.

Company Dinner

The salad and dessert can be prepared in the morning and set aside. The main course can be prepared, ready to cook, and set aside. The remaining meal can be put together by following the steps outlined here. Once the preplanning steps have been taken, the entire meal can be assembled and served in about 30 minutes.

Spicy Madrilene
Chicken St. George
Rice Pilaf
Endive and Watercress Salad
Dinner Rolls
Party Fondue

1. Wash and chill salad greens. Refrigerate.
2. Prepare Party Fondue. Cover fruits and cake tightly and set aside. Leave chocolate mixture in small mixing bowl. Cover.
3. Prepare Chicken St. George up to step 3.
4. Prepare Rice Pilaf. Let stand.
5. Prepare Spicy Madrilene. Serve.
6. Finish Chicken St. George. Toss salad. Serve chicken, rice, and salad.
7. Heat rolls. Serve.
8. Reheat fondue sauce. Serve dessert.

Spicy Madrilene   4 cups

2 cans (13 ounces each) consomme madrilene

Dash hot pepper sauce
¼ cup dry sherry

1. Combine consommé and hot pepper sauce in a 1-quart casserole or 1-quart glass measure. Cook on HIGH for 3 to 4 minutes to 160°F.
2. Stir in sherry.
3. Serve immediately in warmed soup cups.

Chicken St. George   4 servings

2 large, boneless, skinless chicken breasts
¼ cup butter or margarine
Salt and pepper

4 thin slices boiled ham
4 thin slices Mozzarella cheese
2 tablespoons grated Parmesan cheese

1. Cut each chicken breast in half. Place between two pieces of waxed paper and flatten with a mallet or the broad side of a cleaver.
2. Place butter in an 8-inch square glass baking dish. Cook on HIGH for 30 seconds, or until melted. Add chicken breasts, turning once to coat with butter. Cook, covered, on HIGH for 10 to 11 minutes, or just until chicken is tender.
3. Sprinkle with salt and pepper to taste. Top each piece of chicken with a slice of ham, a slice of Mozzarella cheese, and a sprinkling of Parmesan cheese. Cook on HIGH for 2 to 3 minutes, or until cheese melts.
4. Let stand 3 to 4 minutes before serving.

Rice Pilaf   4 servings

¼ cup butter or margarine
1 small onion, finely chopped
½ cup diced celery

2 cups chicken broth
1 cup long grain rice
2 tablespoons chopped parsley

1. Combine butter, onion, and celery in a 1½-quart glass baking dish. Cook, covered, on HIGH for 3 to 4 minutes. Add chicken broth. Cook, covered, on HIGH for 3 to 4 minutes, or until broth comes to a boil.
2. Stir in rice. Cook, covered, on SIMMER for 15 to 18 minutes, or just until rice is barely tender.
3. Let stand, covered, 3 to 5 minutes before serving. Stir in parsley just before serving.

Party Fondue   4 servings

1 package (12 ounces) semisweet chocolate bits
2 squares (1 ounce each) unsweetened chocolate
1 cup heavy cream
3 tablespoons brandy

2 bananas, cut into chunks
1 pint fresh strawberries, washed and drained
Fresh or canned pineapple cubes
Pound cake, cut into 1-inch cubes

1. Combine chocolate bits and unsweetened chocolate in a small glass mixing bowl. Cook, covered, on BAKE for 4 to 5 minutes, or until chocolate melts. Watch carefully during last minute to prevent overcooking.
2. Stir in cream with a wire whisk to make a smooth paste. Cook, covered, on BAKE for 1 to 1½ minutes, or until hot. Stir in brandy.
3. Serve from a fondue pot with fruit and cake.
4. To reheat: If fondue has been standing at room temperature, cook, covered, on BAKE for 1½ to 2 minutes. If refrigerated, for 3 to 3½ minutes.

Family Dinner

This is a menu that is a joy for the cook. The lasagna can be prepared in the morning, along with the greens for the salad and the garlic bread. The broccoli can be prepared and let stand. The preparation of the melon and dessert are last minute items, so cooking—just before supper—can be done in about 30 minutes. During heating time for the lasagna, the first course can be served and preparations made for dessert.

Melon with Prosciutto Ham
Lasagna
Tangy Broccoli
Tossed Green Salad
Garlic Bread
Baked Rum Bananas

1. Prepare greens for salad and refrigerate.
2. Prepare garlic bread through step 2. Set aside.
3. Make lasagna. Let stand.
4. Cook broccoli and sauce.
5. Cut melon into six wedges and cover each with a slice of prosciutto ham. Serve.
6. Toss salad. Heat garlic bread.
7. Serve lasagna, broccoli, and salad. Serve garlic bread.
8. Prepare bananas and serve.

Lasagna

6 to 8 servings

1 pound lean ground beef
1 tablespoon salad oil
2 cloves garlic, minced
2 tablespoons chopped parsley
½ teaspoon oregano
½ teaspoon basil
1 can (15 ounces) tomato sauce
1 can (6 ounces) tomato paste
½ cup water
8 ounces lasagna noodles, cooked,* divided
1½ cups ricotta or cottage cheese, divided
½ pound grated Mozzarella cheese, divided
½ to ¾ cup grated Parmesan cheese, divided

1. Combine crumbled beef, salad oil, and garlic in a 1½-quart casserole or baking dish. Cook, covered, on HIGH for 3 minutes. Break up meat with a fork. Cook, covered, on HIGH for 6 more minutes. Add parsley, oregano, basil, tomato sauce, tomato paste, and water. Stir well. Cook, covered, on HIGH for 4 minutes, then cook on SIMMER for 10 minutes.
2. Pour ½ cup of the sauce in the bottom of a 12- by 7-inch glass baking dish. Top with half the cooked lasagna noodles, half the ricotta cheese, half the Mozzarella cheese, half the meat sauce, and half the Parmesan cheese. Make a second layer with remaining ingredients, ending with Parmesan cheese.
3. Cook, covered, on ROAST for 20 to 25 minutes, or until bubbling.
4. Let stand, covered, 10 minutes.
5. Cut into squares to serve.

*Follow Microwave instructions on pasta chart for cooking noodles.

Tangy Broccoli

6 servings

1½ to 2 pounds fresh broccoli
¼ cup butter or margarine
⅓ cup sliced green onions, including green tops
1 jar (4 ounces) chopped pimiento
1 tablespoon lemon juice
½ teaspoon salt
⅛ teaspoon pepper
Pinch of oregano

1. Cook broccoli according to directions in vegetable cooking chart. Cover and let stand.
2. Combine butter and onions in a 2-cup glass measure. Cook on HIGH for 2 minutes, or until onions are transparent. Stir in pimiento, lemon juice, salt, pepper, and oregano. Cook on HIGH for 30 seconds.
3. Pour over cooked, drained broccoli.

Garlic Bread

1 loaf

½ cup butter or margarine
1 teaspoon garlic powder
1 loaf (1 pound) French bread

1. Place butter in a 2-cup glass measure. Cook on ROAST for 1½ minutes, or until butter is melted. Stir in garlic powder.
2. Cut French bread in half. Slice each half into 1-inch thick slices. Brush both sides

of each slice with butter mixture. Reshape each half loaf. Wrap in waxed paper or paper towels.

3. Cook on REHEAT for 1 to 1½ minutes, or until warmed.

Baked Rum Bananas

6 servings

½ cup butter
½ cup firmly packed dark brown sugar
6 small ripe bananas

¼ cup dark rum
6 scoops hard-frozen vanilla ice cream

1. Combine butter and sugar in an 8-inch square glass baking dish. Cook on HIGH for 2 minutes. Stir well. Cook on HIGH for 1 minute, or until mixture bubbles.
2. Peel bananas and cut in half lengthwise. Arrange in butter mixture. Cook on HIGH for 45 seconds. Turn bananas over and pour rum over top. Cook on HIGH for 45 seconds.
3. Place ice cream in dessert dishes. Place two banana halves on either side of each ice cream scoop. Top with syrup.
4. Serve immediately

Ladies' Bridge Day Luncheon

Before the ladies arrive, this meal can be prepared and wrapped up. Dessert and salad can be made ready and set aside. Coquilles St. Jacques can be prepared through step 3. The mixture can then be quickly heated as the salad is tossed. A toast to the hostess!

Coquilles St. Jacques
Wilted Spinach Salad
Hot Rolls
Lemon Squares

1. Prepare Spinach Salad recipe through step 2. Wash spinach and chill.
2. Make Lemon Squares. Set aside.
3. Prepare Coquilles St. Jacques.
4. Finish Spinach Salad. Heat rolls. Serve Coquilles St. Jacques, salad, and rolls.
5. Serve Lemon Squares.

Coquilles St. Jacques

4 servings

¼ cup butter or margarine
¼ cup chopped celery
1 cup sliced fresh mushrooms or 1 jar (4 ounces) sliced mushrooms, drained
2 medium green onions, thinly sliced
2 tablespoons all-purpose flour
½ teaspoon salt
⅛ teaspoon pepper
½ cup dry white wine

1 pound fresh or frozen scallops, thawed
¼ cup light cream
1 egg yolk
2 tablespoons butter or margarine, melted
2 tablespoons dry bread crumbs
2 tablespoons grated Parmesan cheese

1. Place butter, celery, mushrooms, and onions in a 2-quart casserole or baking dish. Cook, uncovered, on HIGH for 3 minutes, stirring once during cooking. Stir in flour, salt, pepper, and wine, and blend well.
2. Rinse scallops. If large, cut in halves or quarters. Add to mixture in casserole. Cook, covered, on HIGH for 5 to 6 minutes, until mixture boils and thickens.

3. Beat together cream and egg yolk. Stir into scallop mixture. Cook, uncovered, on BAKE for 3 to 4 minutes, stirring once during cooking.
4. Spoon mixture into 4 scallop shells or individual serving dishes that hold about 1 cup each.
5. Melt butter on HIGH for 1 minute in 1-cup glass measure. Add bread crumbs. Sprinkle over top of scallop mixture. Sprinkle cheese over top. Cook, uncovered, on BAKE for 2 minutes, or until cheese is melted and mixture bubbles.
6. Serve immediately.

Wilted Spinach Salad
4 servings

6 slices bacon	1 pound fresh spinach, washed
¼ teaspoon pepper	and torn into bite-size pieces
2 tablespoons sugar	2 green onions, thinly sliced
⅓ cup wine vinegar	1 hard-cooked egg, finely
2 tablespoons water	chopped

1. Arrange bacon slices in a single layer in a 12- by 7-inch glass baking dish. Cook on HIGH for 4 to 5 minutes, or until bacon is crisp.
2. Drain and reserve bacon fat in 2-cup glass measure. Crumble bacon and set aside. Add pepper, sugar, vinegar, and water to fat. Cook on HIGH for 2 to 2½ minutes, or until hot.
3. Place spinach in a salad bowl. Pour hot bacon-fat mixture over spinach. Add crumbled bacon, onion, and egg.
4. Toss lightly and serve at once.

Lemon Squares
6 to 9 servings

1 cup all-purpose flour	1 tablespoon all-purpose flour
½ cup butter or margarine, softened	½ teaspoon baking powder
	1 cup sugar
¼ cup confectioners sugar	Grated peel and juice of 2 lemons
2 eggs	

1. Mix together flour, butter, and confectioners sugar. When well blended, press lightly into 8-inch square glass baking dish. Cook on HIGH for 3 to 4 minutes, or until firm.
2. Beat eggs until light and thick. Stir in flour, baking powder, and sugar to make a smooth mixture. Add lemon peel and juice. Mix well. Pour over baked layer. Cook on HIGH for 5 minutes, or until slightly set.
3. Cut into squares and serve warm.

Convenience Foods

Most precooked foods can be heated in the microwave oven on the REHEAT setting. Precooked foods with cheese, sour cream, or eggs should be prepared using the ROAST setting. Use glass baking dishes or casseroles, glass measuring cups, paper plates or napkins, pottery soup bowls or plates to make your convenience foods even more convenient by using the microwave oven.

Convenience Foods—How to Heat, Defrost, or Cook

1. Precooked foods with cheese, sour cream, or eggs should be prepared on ROAST.
2. Most other precooked foods can be heated on REHEAT.
3. Large containers of foods should be started on DEFROST and then heated on either ROAST or REHEAT.
4. Use glass baking dishes, casseroles, glass measuring cups, paper plates, paper napkins, serving bowls, soup bowls, or plates without metal or silver trim. Dishes can be covered with waxed paper or plastic wrap.
5. TV dinners and many frozen cooked foods can be reheated in the aluminum foil tray, providing the tray is no more than ¾-inch deep. Remove cover and cover loosely with plastic wrap before heating.
6. Food in aluminum foil containers deeper than ¾ inch should be turned out into a glass or pottery container before heating.
7. Canned vegetables can be turned into a small bowl and heated to 150°F. with the use of the probe.

Product and Size	Container	Setting and Time	Special Notes
Appetizers			
Egg rolls, 6-ounce container	Place on plate on paper napkin.	ROAST, 2½ to 3 minutes	Stand 1 minute.
Swiss cheese fondue	glass bowl	ROAST, 5 to 6 minutes	Slit pouch. Press out into bowl for serving.
Beverages			
Milk	glass, cup, or measuring cup	ROAST	
One 8-ounce		2¾ to 3 minutes	
Two 8-ounce		3¼ to 3½ minutes	
Water	glass, cup, or measuring cup	HIGH	
One 8-ounce		1½ to 3 minutes	
Two 8-ounce		3 to 3½ minutes	
To reheat coffee	glass or cup	HIGH	
One 8-ounce		1¼ to 1½ minutes	
Two 8-ounce		2 to 2½ minutes	
Bread			
1 slice, frozen	paper towel or napkin	DEFROST, 15 to 20 seconds	
1 loaf	in package	DEFROST, 2 to 3 minutes	Stand 3 minutes. Remove wire twister before defrosting.
Buns and rolls (hot dog, dinner, hamburger), room temperature	paper plate, towel, or napkin	REHEAT	Add 5 seconds if frozen.
1		5 to 10 seconds	
2		10 to 15 seconds	
4		15 to 20 seconds	
6		20 to 25 seconds	

Product and Size	Container	Setting and Time	Special Notes
Sweet rolls, coffee-cakes, muffins, room temperature	paper plate, towel, or napkin	REHEAT	Add 5 seconds if frozen.
1		15 to 20 seconds	
2		25 to 30 seconds	
4		35 to 45 seconds	
6		45 to 60 seconds	
Whole coffeecake		REHEAT	
Room temperature		1 to 1½ minutes	
Frozen		1½ to 2 minutes	
French bread, 1 pound		REHEAT	
Room temperature		30 to 40 seconds	
Frozen		1½ to 2 minutes	

Cakes

Product and Size	Container	Setting and Time	Special Notes
12 to 17 ounces	paper plate, towel, or napkin	DEFROST, 2 to 3 minutes	Remove from paper carton before defrosting.

Cookies

Product and Size	Container	Setting and Time	Special Notes
Brownies or bar cookies, 12 or 13 ounces	original tray, remove lid	DEFROST, 1½ to 2 minutes	
Cookies	paper plate, towel, or napkin	DEFROST	
1		15 to 20 seconds	
2		30 to 35 seconds	
6		50 to 60 seconds	

Fish

Product and Size	Container	Setting and Time	Special Notes
Fish patties, breaded, frozen	serving plate	REHEAT	
1		1 to 2 minutes	
2		3 to 3½ minutes	
6		5 to 6 minutes	
Fish sticks, breaded, frozen	serving plate	REHEAT	
4		2 to 3 minutes	
8		3½ to 4½ minutes	
1 pound	baking dish	HIGH, 9 to 11 minutes	

Main Dishes

Product and Size	Container	Setting and Time	Special Notes
Creamed chicken, 6½ ounces	baking dish	REHEAT, 7 to 9 minutes	Slit plastic pouch. Let stand. Turn out and stir.
Main dish mixes:			
Macaroni or noodles, 1 pound hamburger	2½- to 3-quart covered casserole	HIGH, 5 minutes HIGH, 5 minutes SIMMER, 10 to 14 minutes	To brown hamburger. After adding dry mix from carton. Let stand, covered, 5 minutes.
Rice (7 ounces), 1 pound hamburger	2½- to 3-quart covered casserole	HIGH, 5 minutes HIGH, 5 minutes	Same as above.

Product and Size	Container	Setting and Time	Special Notes
Lasagna, 21 to 26 ounces	Remove from container. Place in 2-quart glass covered baking dish.	ROAST, 20 to 22 minutes	Let stand, covered, 5 minutes.
Spaghetti sauce:			
32 ounces, canned	1½-quart covered casserole	REHEAT, 8 to 10 minutes	Stir during heating.
1 pint, frozen	1-quart covered casserole	DEFROST, 1 to 2 minutes REHEAT, 6 to 8 minutes	Heat in container on DE-FROST. Turn out into casserole. Cook on RE-HEAT. Stir and heat until hot.
TV Dinners	Leave in foil tray. Cover with waxed paper.	REHEAT	Foil tray should not be more than ¾ inch deep.
One 8- to 11-ounce tray		8 to 10 minutes	
Two 8- to 11-ounce trays		14 to 16 minutes	

Poultry

Product and Size	Container	Setting and Time	Special Notes
Fried chicken, breaded pieces, frozen	Place on plate.	REHEAT	
1 piece		2½ to 3½ minutes	
2 pieces		3⅓ to 4½ minutes	
4 pieces		5 minutes	
2 to 3 pounds, frozen	Remove plastic wrap. Place in glass baking dish.	10 to 12 minutes	
Chicken Kiev, frozen	Remove plastic wrap. Place on plate.		Chicken Kiev needs two temperatures for best results.
1 piece		DEFROST, 4 to 5 minutes HIGH, 2½ to 3 minutes	
2 pieces		DEFROST, 6 to 7 minutes HIGH, 4 to 5 minutes	
Chicken cordon bleu	Remove plastic wrap. Place top side down on plate.		Cook on defrost. Turn top side up and continue cooking.
1 piece		DEFROST, 4 to 5 minutes HIGH, 2½ to 3 minutes	
2 pieces		DEFROST, 6 to 7 minutes ROAST, 4 minutes (turn plate) ROAST, 3 minutes	

Product and Size	Container	Setting and Time	Special Notes
Rice			
1 cup cooked, refrigerated	covered dish or bowl	REHEAT, ½ to 2 minutes	Stand 2 minutes.
2 cups		REHEAT, 3 to 4 minutes	
Pouch, 11 ounces	glass baking dish	REHEAT, 6 to 7 minutes	Slit pouch before cooking. Stand. Remove from pouch and stir lightly.
Soups			
Canned soups	1½-quart bowl	HIGH, 5½ minutes	Pour canned soup into 1½-quart casserole. Add milk or water. Heat to 150°F. to 160°F. with probe.
1 bowl	soup bowl without metal trim	REHEAT, 3⅓ to 4 minutes	
2 bowls		REHEAT, 7 to 8 minutes	
Vegetables			
Au Gratin, 1½ ounces	Place in loaf dish. Cover with waxed paper.	ROAST, 10 to 12 minutes	Stand, covered, 3 minutes.
Potatoes, baked, stuffed	1-quart casserole	ROAST, 10 to 12 minutes	Cover with waxed paper first half of cooking time. Remove last 5 minutes of cooking time.
Spinach soufflé, 11½ ounces	Uncover dish. Cover with waxed paper.	DEFROST, 5 minutes REHEAT, 10 minutes	Let stand 3 minutes.
Mixed Vegetables	1-quart covered casserole	HIGH, 7 to 8 minutes	Stir halfway through.

INDEX